Interacting Binary Stars

S381
The Energetic Universe
3

Ulrich Kolb

S381 COURSE TEAM

Course Team Chair	Dr Andrew Norton
Authors	Dr Carole Haswell, Dr Ulrich Kolb, Dr Sean Ryan
Reader	Dr Mark Jones
Course Manager	Gillian Knight
Editor	Dr Rebecca Graham
Course Secretaries	Valerie Cliff, Tracey Moore
Software Designer	Dr Will Rawes
Multimedia Producer	Dr Kate Bradshaw
Graphic Designer	Debbie Crouch
Graphic Artist	Pam Owen
Picture Researcher	Deana Plummer
Consultants	Dr Katherine Blundell, Oxford University Dr Atsunori Yonehara, Tsukuba University
Block Readers	Dr Christopher Tout, Cambridge University Professor Andrew King, Leicester University Dr Christine Done, Durham University
Course Assessor	Professor Michael Merrifield, Nottingham University

This publication forms part of an Open University course S381 *The Energetic Universe*. The complete list of texts which make up this course can be found on the back cover. Details of this and other Open University courses can be obtained from the Course Information and Advice Centre, PO Box 724, The Open University, Milton Keynes MK7 6ZS, United Kingdom: tel. +44 (0)1908 653231, e-mail ces-gen@open.ac.uk

Alternatively, you may visit the Open University website at http://www.open.ac.uk where you can learn more about the wide range of courses and packs offered at all levels by the Open University.

To purchase this publication or other components of Open University courses, contact Open University Worldwide Ltd, Walton Hall, Milton Keynes MK7 6AA, United Kingdom: tel. +44 (0)1908 858785; fax +44 (0)1908 858787; e-mail ouwenq@open.ac.uk; website http://www.ouw.co.uk

The Open University
Walton Hall, Milton Keynes
MK7 6AA

First published 2002

Edited, designed and typeset by The Open University.

Printed and bound in the United Kingdom by Bath Press, Bath.

ISBN 0 7492 9765 4

1.1

CONTENTS

AIM

The aim of Block 3 is to achieve a deeper understanding of the physical principles that govern the processes of mass transfer and accretion in interacting binary stars.

LEARNING OUTCOMES

This block provides opportunities for students to develop and demonstrate the following learning outcomes:

1 A familiarity with the terminology which is used to describe the properties and behaviour of interacting binary stars.

2 Knowledge of how the physical properties of binary stars and of the accretion flow in these systems are measured.

3 An understanding of the physical mechanisms that underlie binary star evolution and accretion in binary stars, and the ability to describe these mechanisms quantitatively.

4 An understanding of basic concepts of hydrodynamics, thermodynamics and plasma physics that are of relevance to astrophysics.

5 The ability to derive and manipulate quantitative theoretical models of physical processes and to derive physical estimates.

6 An appreciation of the interplay between theoretical modelling and observation or experiment.

7 The ability to organize and clearly present relevant information in response to defined tasks, including the expression of mathematical and scientific concepts using clear, concise and correct scientific prose.

8 The ability to learn from a variety of sources and media including books and journal articles which are not specifically written for an undergraduate audience.

9 The ability to evaluate and synthesize information from a variety of sources and media.

10 The ability to search for and download relevant information from the World Wide Web.

11 The ability to use spreadsheets to model physical processes and present the results graphically.

INTRODUCTION

So far you have been studying the life and death of single stars. In this block we shall consider stellar couples – binary stars. The Sun has no stellar companion: the most massive planet in the Solar System, Jupiter, is not dense and hot enough to initiate hydrogen burning in its core. But it is perfectly feasible that two stars are born in such close proximity to each other that they stay gravitationally bound, orbiting about their common centre of mass. The complex, sometimes violent, phenomena that arise when two stars live close to each other are the subject of this block.

The foundation of this block is the first half of the textbook *Accretion Power in Astrophysics* by Juhan Frank, Andrew King and Derek Raine – which we simply call FKR in the following text. FKR is the most advanced of the three textbooks you have received for this course – and this is not just because of the large number of equations it contains. The bigger challenge is likely to be FKR's leisurely use of difficult concepts and techniques. As if this were not enough, FKR also consistently uses cgs units rather than SI units. So be prepared for a minor culture shock. But relax: the whole point of this Study Guide is to make the textbook more accessible. You will have to read only a selection of paragraphs, and make use of only a fraction of the equations in the textbook. Although you are welcome to read the rest of the book at your leisure, you should remember that this is *not* required for a successful completion of the course. Do not allow FKR to distract you from all the other fascinating activities in this Study Guide and in the other blocks.

The main activities in this block are readings of selected paragraphs in FKR. In most cases the Study Guide provides detailed comments and further explanations to these readings immediately after the activity. So even if you think you are stuck somewhere in the textbook and feel you have no clue what it means – don't despair, just keep on reading. You can always revisit the problem points in the textbook after you've worked through the commentary in the Study Guide. It is for this reason that you should work through all examples and questions where they appear. They are an integral part of our attempts to assist you in using FKR as a textbook. Some of the worked examples introduce new concepts and ideas. You are welcome to attempt the solution on your own, but please do study the model answer carefully. For some of the questions you will need the actual value of physical constants; if these are not quoted in the text of the question you will find them in the Appendix. Do not try to memorize all the equations you will encounter in this block – there are way too many of them. As you will hopefully discover, equations are not just a combination of symbols, they convey a piece of physics, and it is the physical meaning we wish you to concentrate on. If you do that you will always be able to find the equation you need in the Study Guide, the textbook, or in the list of equations provided with this course.

The block consists of nine main sections, and by and large you should complete one section per study week. The only exception to this rule is Section 3 and Section 4, which taken together should be completed in two weeks. Section 4 is the core part of this block and provides the theoretical foundation for the understanding of accretion discs. It is likely that you will need more than one week to work through Section 4, so you should aim for an early finish of Section 3, to give yourself a head start on Section 4.

Section 1 — world of interacting binary stars

Section 2 — stars feel on constraints when they live as a binary — can be mass transfer between one star accretes at expense of the other

Section 3-5 physics of accretion — setting out with fundamental physical laws + concepts, + then developing simple ideas + models of how accretion goes

Section 6-7 test models + their predictions against observation.

Section 8 — most advanced view on the physics behind the spectacular outbursts in interacting stars

We begin the block fairly gently with a travel guide to the world of interacting binary stars (Section 1). This is backed up by a more quantitative discussion of the constraints stars feel when they live in a binary. An important discovery will be that there can be mass transfer between the two component stars, leading one star to accrete material from the other (Section 2). Much of the rest of the block is devoted to the fascinating physics of accretion. A characteristic of the way the material in this block is presented is that we set out with fundamental physical laws and concepts to develop simple ideas and models of how accretion proceeds (Sections 3, 4 and 5). Only then do we test the models and their predictions against observations (mainly in Sections 6 and 7). The most advanced view on the physics behind the spectacular outbursts in interacting binary stars will be presented in Section 8. This is also conceptually the most difficult part of the block, but as it is a long way off, there is no need to worry yet – and in fact, no need to worry at all.

1 BINARY STARS

1.1 Stars come in pairs

Before we embark on our extended tour of the world of **binary stars** we should convince ourselves that we are not wasting our time by dealing with just one of the many rare stellar oddities. Once we understand binary stars, do we then have to consider triple stars, after that quadruple stars, then perhaps quintuple stars, and so on? Fortunately, the answer is no – for a very simple reason. Unlike binary stars, gravitationally bound several-body systems are usually unstable, and do not stay bound for long. Even if a triple star manages to form, the third star would soon be lost into space, leaving a stable binary behind (see the example shown in Figure 1). The same is true for bound multiple stars with more than three components. But there is an important exception to this rule. Multiple stars can survive for a long time if they form a **hierarchical binary**. These consist of two main components orbiting each other. Any of these components can itself be a pair of stars, but the separation of stars in the pair must be much smaller than the separation of the main components. A star of the pair could itself be a binary, with a yet smaller separation (see Figure 2 overleaf). A famous example is the **visual binary star** ε Lyrae, in the constellation Lyra (the lyre) (Figure 3 overleaf). With some effort it can be seen to be a double star with the naked eye, but it is much easier to see with binoculars. With a telescope and moderate magnification, each of the two stars resolves into two very close components. So ε Lyrae is a hierarchical quadruple star (see also Figure 4 overleaf).

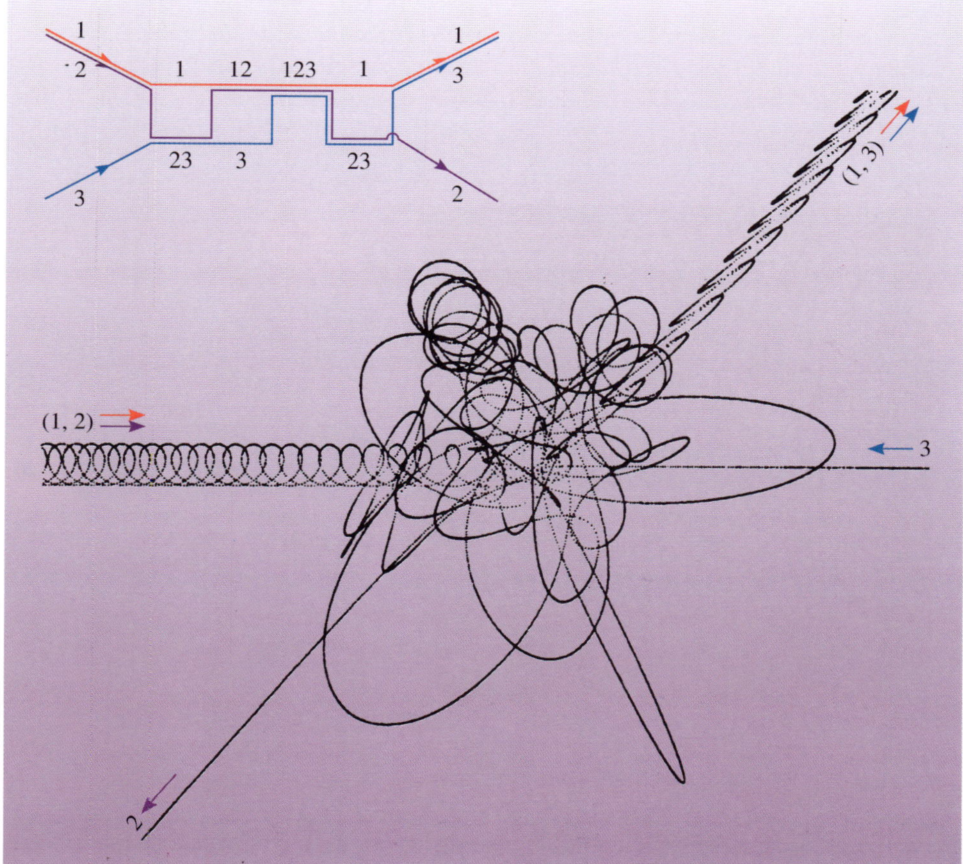

Figure 1 Trajectories of a binary star colliding with a single star. The unstable triple-star configuration survives only for a short time. The outcome of the collision is a partner swap. Star 2 escapes, leaving stars 1 and 3 behind as a newly formed binary. The schematic diagram (top left) shows the hierarchical structure of the triple star during the collision process, i.e. indicates which of the stars form a transient pair.

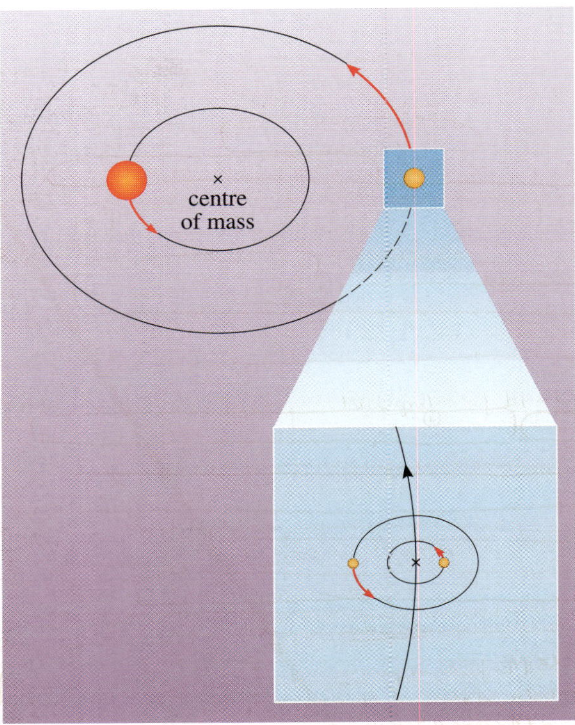

Figure 2 Schematic diagram showing a hierarchical binary system that consists of three stars.

[Handwritten margin notes: stars seem to prefer being a binary rather than single. binary fraction is difficult to measure — some of binary characteristics work against discovering them]

Stars seem to prefer life as a member of a binary system over solitude. Statistical counts suggest that more stars reside in binaries than there are single stars. The true incidence of binarity, the **binary fraction**, is difficult to measure because some of the characteristics of binaries work against their discovery, for example, wide orbits with correspondingly

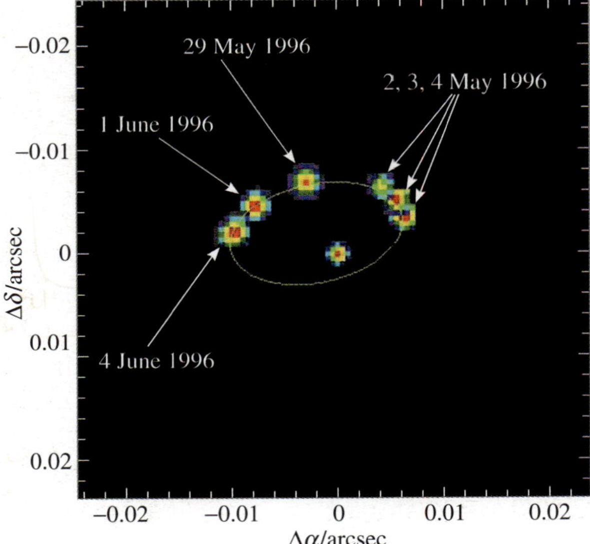

Figure 4 Orbit of component A of the visual binary ζ Ursae Majoris (Mizar). Mizar is a well-known hierarchical binary, with components A and B separated by about 14 arcsec. This composite image of the orbital motion of the two much closer stars that make up component A was obtained using an **optical interferometer**, which achieves a much higher angular resolution than a telescope on its own. Component B is off the page.

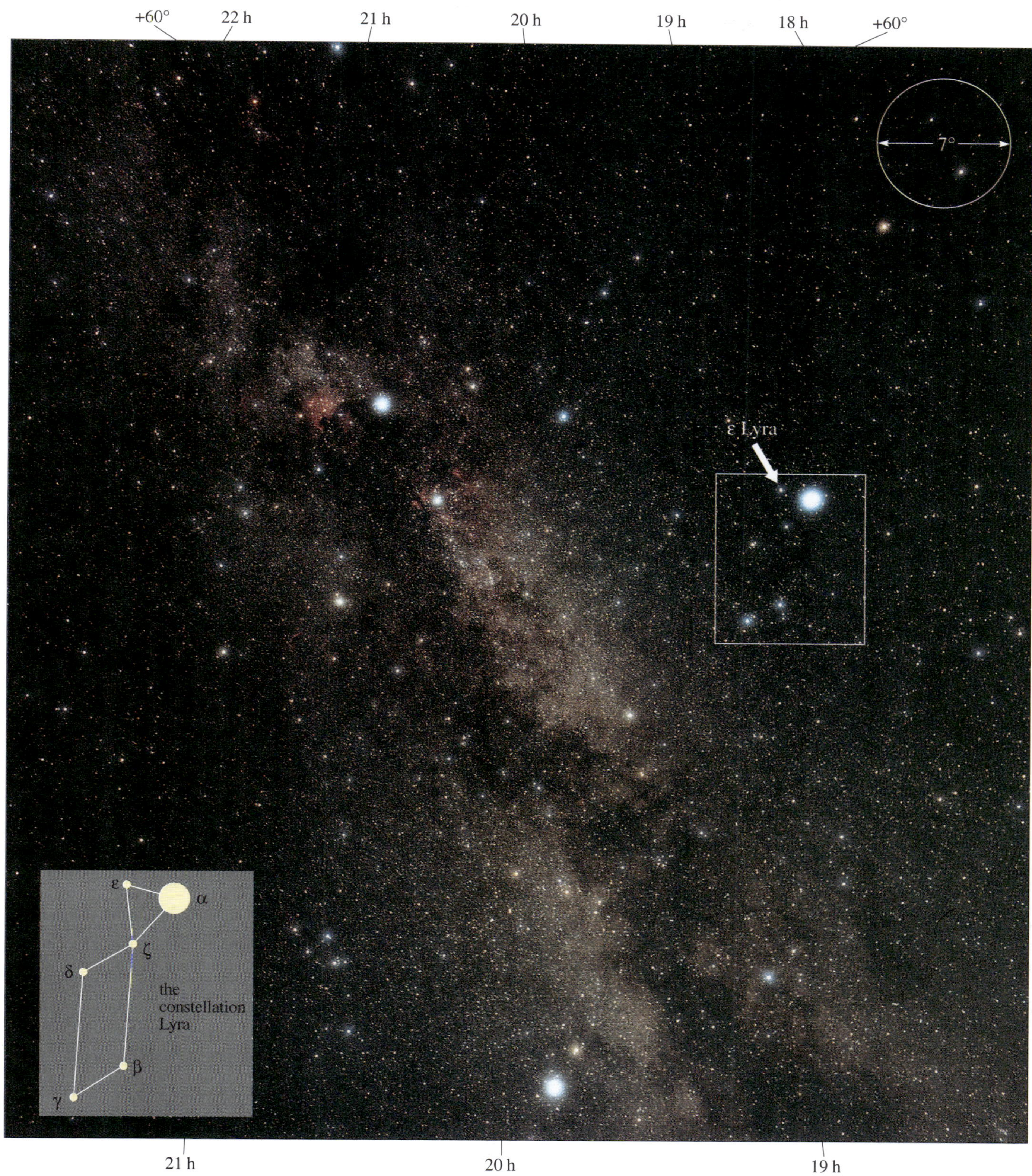

Figure 3 The night sky showing the Cygnus region of the Milky Way. The constellation of Lyra is marked. The arrow indicates ε Lyrae.

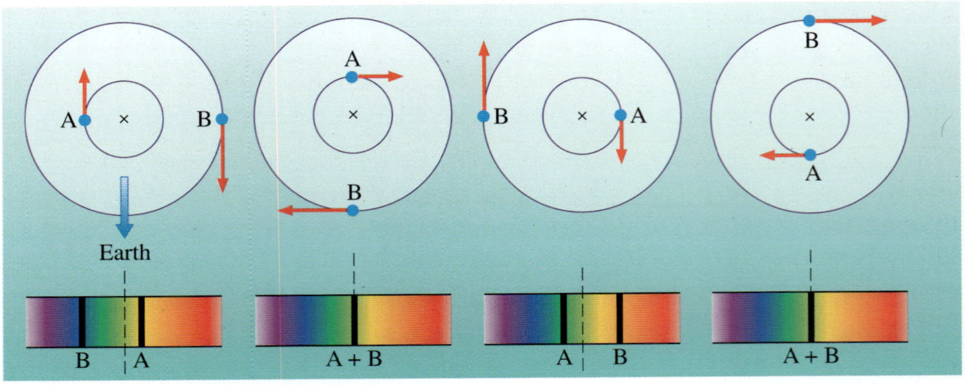

Figure 5 A spectroscopic binary cannot be resolved in telescopes, but reveals its identity through the periodic Doppler shift of spectral lines.

rather slow orbital velocities. Except for the few visual binaries, the binary nature of a star – a point-like light source even when observed with the largest telescopes – is usually recognized through a periodic Doppler shift of the spectral lines of at least one of the components due to the orbital motion. If this is the case the system is a **spectroscopic binary** (Figure 5). Recall from Block 1 (Section 4.3.2) that the velocity of a star with mass m in a circular orbit with radius r around another star with mass M is

$$v = \sqrt{\frac{GM}{r}} \tag{1}$$

If the separation is large, the Doppler shift $\propto v$ is small and difficult to measure (see also Block 1, Section 2.4 and Section 2.10).

■ How does Equation 1 follow from the statement that the centripetal force arises from the gravitational force?

❑ If the mass of the orbiting star is m, then the gravitational force is GMm/r^2, while the centripetal force is mv^2/r (see Block 1, Section 4.3.2). Equating these forces, cancelling m, and rearranging for v reproduces Equation 1. ■

Estimates for the binary fraction range from 20% to more than 90%. This preference for binarity is, of course, no coincidence. As you will recall from Block 2, star formation involves the gravitational collapse and fragmentation of a large cloud of gas and dust. Hence protostars stand a good chance of forming in such close proximity to each other that they stay gravitationally bound. Multiple systems might form as well, but for stability reasons they eventually dissolve to form binaries and single stars, like the ones shown in Figure 2.

The very existence of binary stars is a stroke of luck for astrophysicists. The stellar companion effectively constitutes a *measuring device*. Stellar parameters that otherwise are accessible only in a rather indirect way, or inaccessible altogether, suddenly become measurable. A prime example is the mass of a star. We can *weigh* the stars by observing the orbital motion. In eclipsing binaries (see, for example, Figure 48 in Block 1) it is sometimes even possible to measure directly the radii of the two stars.

The presence of a companion can also complicate, indeed alter, the life of a star. New phenomena arise if the orbital separation is small enough that the two stars physically interact, e.g. through mass flowing from one star to the other. This is called **mass transfer**. If the mass gainer is an evolved, old star it might rejuvenate when it acquires unspoilt, hydrogen-rich material. Conversely, once the mass donor has lost a significant amount of material it may look much older than a star of its current mass usually would. We shall come back to this so-called *Algol paradox* shortly.

Close, interacting binaries host rather spectacular events when the mass gainer is a compact star, i.e a white dwarf, neutron star or black hole. A star is compact when the mass to radius ratio M/R_* is very small, so that matter leaving the mass donor loses a large amount of gravitational potential energy before it accretes onto the compact star. Most of this energy is used to heat up the material as it falls down towards the compact star. Temperatures in excess of 100 000 K in the case of white dwarfs, and tens of millions of kelvin for neutron stars and black holes are the norm. Stellar matter at this temperature emits electromagnetic waves of very high frequencies, so these binaries are powerful sources of high-energy radiation. The brightest sources in the X-ray sky are in fact accreting neutron star and black hole binaries (see Figure 6). These **X-ray binaries** are so bright that they can be seen even in distant galaxies. Many hundreds of them are known in the Andromeda galaxy M31, the nearest galaxy that is similar to our own.

In the following sections we shall consider interacting binaries, and the physical appearance of compact binaries. We shall study the extreme conditions to which the accreting stellar matter is subjected to in some detail. Our main concern will be the physical principles and mechanisms that govern the process of mass transfer and mass accretion. We aim to develop a physical framework within which the wealth of binary star phenomena find a natural, consistent and, where possible, quantitative explanation.

1.2 Giving and taking

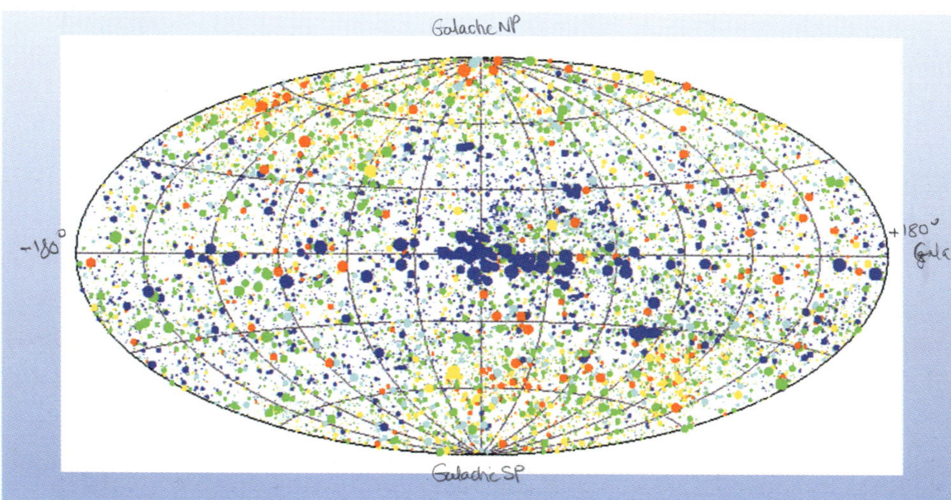

Figure 6 An all-sky map of the X-ray sky as seen by the X-ray satellite ROSAT. The colour of the dots indicates the 'X-ray colour', i.e. the spectral characteristics of the X-ray source. The size of the dots indicates the intensity of the emitted X-rays. The celestial sphere has been mapped into the plane of the page such that the Galactic equator (the band of the Milky Way on the sky) appears as the horizontal line in the middle of the diagram. Along the equator the Galactic longitude changes from −180° at the left to +180° at the right. The Galactic Centre is in the middle. The vertical line in the middle joins the Galactic North Pole (top) with the Galactic South Pole (bottom).

Consider two stars of different masses that orbit each other. Both stars will expand during the course of their evolution. The more massive star evolves faster (see Block 2, Section 5.1) and hence its radius increase will be faster and more pronounced. If the orbit is not too wide, this star will eventually grow so large that its outermost layers experience a stronger gravitational pull from the second star than from their parent star. As a consequence, stellar matter – stellar plasma – flows towards the companion star. It is clear that this mass transfer must occur from the vicinity of the point on the star closest to its companion, the so-called **inner Lagrangian point**, or L_1 point for short (see Figure 7). As we shall see below, matter leaving the ~~primary~~ donor star at the L_1 point forms a geometrically thin stream of gas. The stream accelerates towards the companion, gaining kinetic energy as it loses gravitational potential energy.

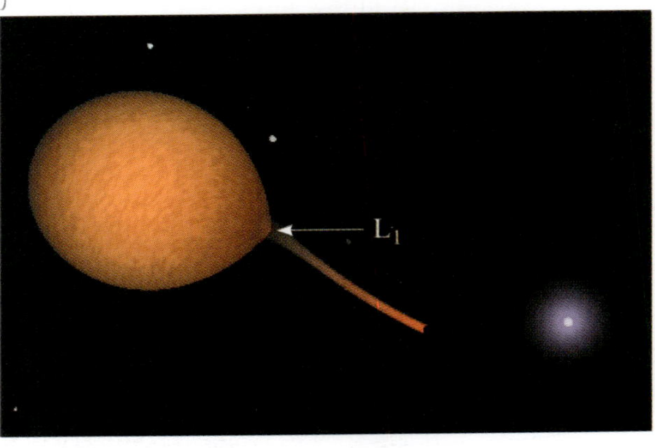

Figure 7 Mass leaving the star in the vicinity of the L_1 point.

If the companion is large enough the stream will hit its surface. Upon impact the kinetic energy is converted effectively instantly into thermal energy, i.e. the plasma heats up enormously. The plasma flow experiences a **shock**: the plasma temperature and density change almost discontinuously (over a tiny length scale, the width of the shock) when the plasma passes through the position of the shock (Figure 8).

FLUIDS AND STELLAR PLASMA

Liquids and gases are collectively called **fluids**. In this course we are dealing with astrophysical fluids like stellar matter. Stellar matter is a gas and often referred to as stellar plasma or cosmic plasma. A **plasma** is a conducting fluid, (e.g. an ionized gas) whose properties are determined by the existence of ions and electrons. Note that if stellar matter is cool enough these ions and electrons may recombine to form neutral atoms and molecules.

Handwritten margin notes:

m_2 m_1

① both stars will get bigger as they evolve.
② more massive will evolve faster hence R increase will be faster.
③ If orbit is not too large the larger star will eventually grow so large that its outermost layers experience a stronger gravitational pull from the other star than from its own star.
④ so stellar matter will flow towards the companion star – it flows from the point closest to the companion (inner Lagrangian point)
⑤ matter leaving donor star at the L_1 point forms a thin stream of gas – the stream accelerates towards companion gain E_{KE} and loses E_{GR}
⑥ if companion large enough stream will hit its surface – upon impact the E_{KE} is converted into thermal E (heats plasma)
⑦ plasma flow experiences a shock – plasma T & ρ change over a tiny length when passes thro' shock position discontinuously
⑧ if the companion star is rather smaller (compact star) the conservation of angular momentum forces stream to swing into an orbit around the compact star – the plasma eventually spread to form a flat disc of gas in the orbital plane centred on the star – an accretion disc

hot subsonic *(less than speed of sound)* settling flow *(slow, hot, dense)*

shock

cool supersonic *(greater than speed of sound)* accretion flow *(cool, fast + thin)*

photosphere

(a)

$z < 1$ $z = D$ $z > D$

z

1.0

0.5

temperature

0.0 0.5 1.0 z/D

shock *(discontinuous change in T, A)*

density

(b)

z

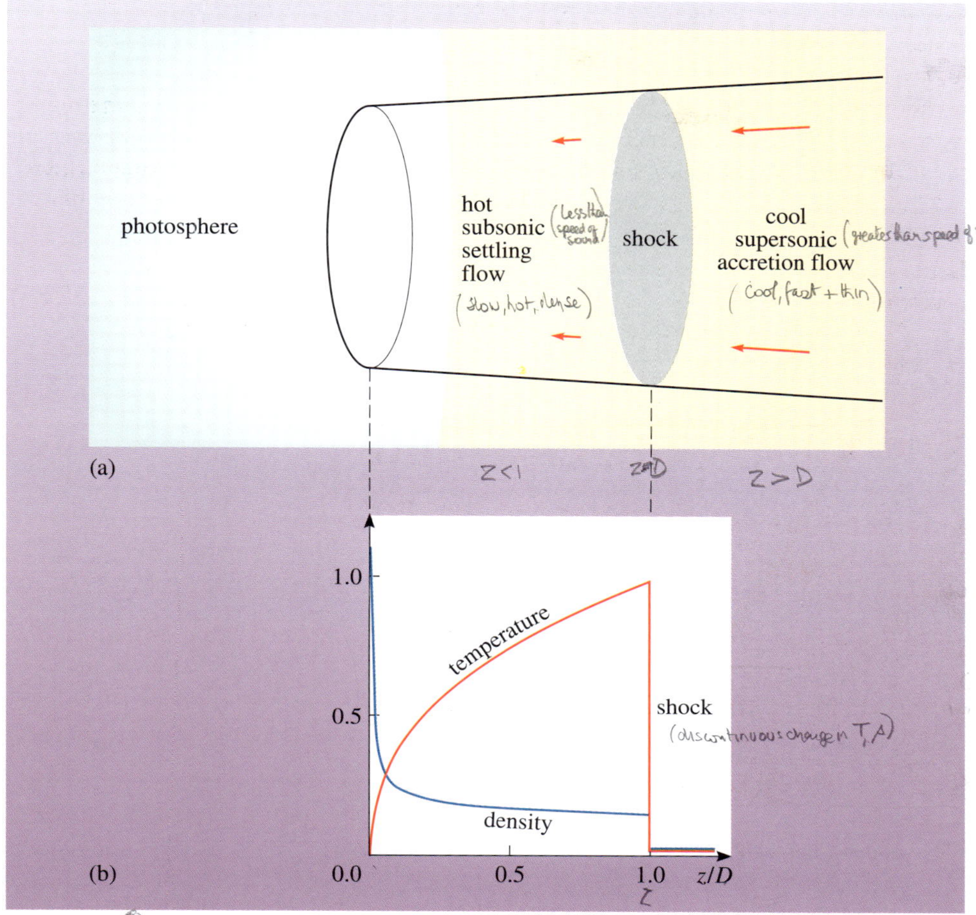

Figure 8 The temperature and density along a plasma flow changes discontinuously at the shock front. The upper panel (a) shows schematically how a plasma stream impacts onto a star. The lower panel (b) shows the temperature (red) and density (blue) of the plasma as a function of height z above the stellar surface. The shock front is at height $z = D$. Above the shock ($z > D$) the plasma is fast, cool and tenuous, while below the shock ($z < D$) it is slow, hot and dense. The flow speed changes from larger than the speed of sound (supersonic) to slower than the speed of sound (subsonic).

If on the other hand the companion is rather small, e.g. a compact star, the conservation of angular momentum forces the stream to swing into an orbit around this star. The plasma eventually spreads to form a flat disc of gas in the orbital plane, centred on the mass gainer. This structure is called an **accretion disc** (see, for example, the cover of this Study Guide). The disc material orbits the companion many times while it maintains a slow drift inwards, thereby losing angular momentum and gravitational potential energy. The liberated energy is converted into heat and eventually emitted in the form of electromagnetic radiation such as visible or UV light, or X-rays. The disc acts as an agent to allow the transferred material to settle gently on the mass accretor. Accretion discs are like machines that extract gravitational potential energy and angular momentum from plasma. (We shall return to the physics of these discs in greater detail later in this block.)

It is instructive to calculate the energy, power and temperatures involved in the process of accretion, and to compare these numbers with the output from other processes. This will be the subject of your first reading of FKR.

If the companion star is small
ⓐ Conservation of angular momentum forces the stream to swing into an orbit around the companion.
ⓑ the plasma spreads out to form a flat disc of gas in the orbital plane centred on the companion — accretion disc
ⓒ disc material orbits companion many times while maintaining a slow inward drift so losing angular momentum + gravitational PE.
ⓓ liberated E is converted into heat + eventually emitted as EMR — visible, UV or x ray.
ⓔ disc acts as the agent that allows the transferred material to settle gently on the companion.
ⓕ accretion discs are like machines that extract E_GR + angular momentum from plasma

Handwritten left-margin notes:

Activity 1
FKR Sections 1.1–1.3
+ Section 1.4 first paragraph

accretion disc luminosity
The energy an accretion disc radiates
per unit time, L_{disc}
For a steady state optically thick, geometrically
thin, infinite disc with a non rotating central
accretor with mass M & radius R_* and
accretion rate $\dot M$
accretion disc $L_{disc} = \dfrac{GM\dot M}{2R_*}$

accretion efficiency (L_{acc})
The accretion luminosity expressed in units
of the rate of accreted mass energy
If L_{acc} denotes accretion luminosity
$\dot M$ mass accretion rate
c speed of light
The accretion efficiency η $\eta = L_{acc}/\dot m c^2$

Cataclysmic variables (CVs)
compact binary systems consisting of a WD
and a low mass ms star in which optical emissions
comes from an accretion disc around WD primary.
Many CVs show regular increases/decreases in L
known as dwarf nova outbursts due to accretion disc
instabilities. Other CVs have much more powerful
classical nova outburst that occur when
enough H has accumulated to allow a temporary
phase of thermonuclear burn on surface of the WD.
CVs with high magnetic fields behave differently.
L_{edd} – the accretion luminosity corresponding
to the situation when an object is accreting matter
at the Edd limit. (This is an approx upper limit
on accretion L since Radiation prevent accretion
at a higher rate)
Edd limit – an upper limit to the L that can be
generated in a system powered by accretion.
Limit arises because photons emitted by accreting
body exert a force on the es in the accreting
material. The es are bound to the ps + other nuclei by
electrostatic forces so the ps feel the force due to
the photons too. At the Edd luminosity the force
exerted by the photons is sufficient to overcome the
inward gravitational attraction.
$L_{edd} = 4\pi G M m_p /\sigma_T$ (M mass of accreting body)
(σ_T is Thomson cross section)
(this assumes spherical accretion, so is only)
(approx for an accretion disc)

$T_b = T$ of a luminous source if it were to
radiate its power in a black body spectrum
– for a star this is its Teffective

T_{virial} – T of a system when the virial
theorem holds is can accrete onto it
(ie there is a balance between thermal +
gravitational potential energy)
Virial theorem holds when for a gravitational
bound system is equilib the total E_{KE} is $-\frac{1}{2} E_{GR}$
$-E_{GR} = 2E_{KE}$ or $E_{KE} = -\frac{1}{2}E_{GR}$

Accretion as a source of energy

Read Sections 1.1–1.3 and the first paragraph of Section 1.4 of FKR.

Remember, if you stumble over the occasional concept or argument you cannot quite follow, just keep on reading. One of the main purposes of this reading is to get you accustomed to the style of FKR. Not everything addressed in these sections is vital for this block. You will be able to identify the points that are important from our explanations below and the follow-up questions.

Keywords: **accretion luminosity, accretion efficiency, cataclysmic variables, Eddington luminosity, Eddington limit, blackbody temperature, virial temperature** ■

NOTATION FOR TIME DERIVATIVES

Often the time derivative of a quantity is denoted by a dot over the symbol representing this quantity. As an example, the mass accretion rate is the rate at which mass is added to an object, i.e. the rate by which the mass M of this object increases. Therefore it can be written as

$$\frac{dM}{dt} \equiv \dot M$$

(handwritten: because matter is falling inward?) and E_{GR} radiated at same rate as liberated

Accretion liberates gravitational potential energy. If this energy is radiated away at the same rate as it is liberated, an object with mass M and radius R_* accreting matter at a rate $dM/dt \,(= \dot M\,)$ shines with the accretion luminosity

(handwritten: energy radiated. energy liberated by infall.)

$$L_{acc} = \frac{GM\dot M}{R_*}$$

(FKR 1.5) (2)

Example 1

(handwritten: $L_{acc} = GM\dot M / R_$)*

Derive the above expression for the accretion luminosity. Consider a small time interval Δt during which a small mass Δm accretes on the star.

Solution

The gravitational potential energy released in the time Δt is just

$$\Delta E_{acc} = \frac{GM\Delta m}{R_*}$$

(FKR 1.1)

This energy is immediately radiated away, so the accretion luminosity must be $L_{acc} = \Delta E_{acc}/\Delta t$. As the mass Δm accretes in the time Δt, the accretion rate is just $\dot M = dM/dt = \Delta m/\Delta t$. So

$$L_{acc} = \frac{GM\Delta m}{R_*}\frac{1}{\Delta t} = \frac{GM}{R_*}\left(\frac{\Delta m}{\Delta t}\right) = \frac{GM\dot M}{R_*} \ \blacksquare$$

(handwritten: accretion L =) ... (handwritten: L_{edd} ✱)

There is an important limit on the rate at which mass can accrete onto a star, the

because ↱ *brought about*

Eddington limit. The radiation generated by accretion exerts an outward force in the radial direction on the incoming material, mediated by the scattering of photons. The dominant scattering targets are free electrons in the accreting material, giving rise to electron scattering or **Thomson scattering.** You have met this process in Block 2 as a major source of opacity in the interior of stars. Just as for opacity or nuclear reactions, the scattering strength is expressed in terms of a cross-section, here the Thomson cross-section σ_T (with units of area). Equating this outward radiation force to the gravitational force gives a maximum luminosity for a gravitationally bound spherical object, the Eddington luminosity

equating the outward radiation force to gravitation force gives max L for a gravitational bound object

$$L_{edd} = \frac{4\pi GM m_p c}{\sigma_T} \approx 1.3 \times 10^{38} \left(\frac{M}{M_\odot} \right) \text{erg s}^{-1} \quad \text{(FKR 1.3 and FKR 1.4)} \quad (3)$$

The Eddington limit of the mass accretion rate arises from the requirement that the accretion luminosity has to be smaller than the Eddington luminosity. When the accretion luminosity just equals the Eddington luminosity, the corresponding mass accretion rate is therefore called the Eddington accretion rate (or simply Eddington rate).

$L_{acc} < L_{edd}$
$L_{acc} = L_{edd}$

incoming material (going to be accreted)

outward flowing radiation generated

Thomson scattering — electron scattering
— important source of opacity in astronomical plasmas especially at high T + low ρ, due to the scattering of photons by free e⁻s as the e⁻ responds to oscillating electric field of photons
e⁻ scatter is independent of λ, depends only on density of free e⁻

Note that σ_T (notation of FKR) is σ_e elsewhere in the course.

$L_{acc} = L_{edd}$, the mass accretion rate is called Eddington accretion rate

Question 1

Explain in words what the difference is between the Eddington luminosity and the accretion luminosity. ■

① L_{edd} = max L of a gravitationally bound object
L_{acc} = rate at which E is liberated when obj accretes matter
If an object shines with L_{edd} the outward radiation P force balances inward gravitational force

Question 2

(a) Calculate the Eddington luminosity for the Sun, and for a neutron star with mass $1M_\odot$.

(b) Calculate the corresponding Eddington accretion rate (assume that the neutron star radius is 10 km). Express these accretion rates in both cgs units and in $M_\odot \text{yr}^{-1}$. ■

A useful concept to emphasize the power of accretion as an energy generator is the accretion efficiency η, defined by (FKR Equation 1.9)

$$L_{acc} = \eta \dot{M} c^2 \quad \left(\text{energy liberated thro' accretion of the blob} \right) \quad (4)$$

In general, the efficiency η expresses the amount of energy gained from matter with mass ΔM, in units of the mass energy $\Delta M c^2$ of that matter. The matter could be, for example, a small amount of plasma, a 'blob' of plasma, in the accretion stream. Equation 4 describes the energy liberated through accretion of the blob.

② $L_{edd,\odot} = 1.3 \times 10^{38} \frac{M_\odot}{M_\odot}$ erg s⁻¹ = 1.3×10^{38} erg s⁻¹ = $3.4 \times 10^{4} L_\odot$
with $L_\odot = 3.83 \times 10^{33}$ erg s⁻¹
(a) NS with mass M_\odot has the same L_{edd}

(b) accretion rate can be calculated because $L_{edd} = L_{acc}$ in this case

η is the Greek letter 'eta'.

$L_{acc} = \frac{GM\dot{M}}{R_*}$

For the Sun $\dot{M}_{edd} = \frac{L_{edd} R_*}{GM_\odot} = 6.8 \times 10^{22}$ g s⁻¹

For NS $\dot{M}_{edd} = (\dot{M}_{edd})_{sun} \times \frac{1 \times 10 \text{km}}{1 R_\odot} = 9.8 \times 10^{17}$ g s⁻¹

$1 M_\odot \text{yr}^{-1} = \frac{1.99 \times 10^{33} \text{g}}{365.25 \times 86400 \text{s}} = 6.3 \times 10^{25}$ g s⁻¹

$\dot{M}_{\odot,edd} = 10^{-3} M_\odot \text{yr}^{-1}$
$\dot{M}_{NS,edd} = 1.6 \times 10^{-8} M_\odot \text{yr}^{-1}$

Question 3

Estimate the efficiency of accretion onto a neutron star. Compare Equation 4 and Equation 2, and use typical parameters of a neutron star, as in FKR. ■

Question 4

(a) Compare the efficiency (= energy gain/mass-energy of input nuclei) of nuclear fusion of hydrogen into helium with the result of the previous question. The mass defect of hydrogen burning is $\Delta m = 4.40 \times 10^{-29}$ kg. (b) Explain why accretion can be regarded as the most efficient energy source in the Universe. ■

extra explanation from erratum
the quoted mass defect refers to the 4 protons involved in H burning only, the overall effective mass defect in H burning is slightly larger (as becomes clear from Section 4.3.2 of Bk2 p88)

③ accretion efficiency $L_{acc} = \eta \dot{M} c^2$, equate this to eq 2 $L_{acc} = \frac{GM\dot{M}}{R_*}$ ∴ $\eta = \frac{GM}{R_* c^2}$
For a NS with mass $M_\odot = 1.99 \times 10^{33}$ g $R_* = 10$ km $\boxed{\eta \approx 0.15}$

④ The mass defect $\Delta m = 4.40 \times 10^{-29}$ kg involved in the fusion of 4p into 1He nucleus translates into energy gain $\Delta E = \Delta m c^2$ per 4ps. Energy input is the mass energy of the 4Ps, $4 m_p c^2$
So the efficiency is gain/input = $\frac{4.40 \times 10^{-29}}{4 m_p c^2} = 0.0066$. The efficiency of the H burn (the most common reaction) is only $\eta = 0.007$, that is 1/10th of H accreted onto a compact star (NS or BH) it liberate 20× more energy than 1 gm of H undergoing nuclear reaction to He. There is no other process in the U which could sustain $M \, \delta E$ for a long time for a macroscopic amount of M.

[handwritten annotations in margins:]

$T = E_{ph}/k = \frac{h}{k}\nu$

$E_{ph} = h\nu$
E_{ph} can be described by a $T = E_{ph}/k$

$E_{ph} = kT$

photon energy $= E_{ph} = h\nu$ given in eV

$1 eV = 1.602 \times 10^{-12} erg$

thermal E of particles of gas with $T = \frac{h}{k}\nu = \frac{E_{ph}}{k}$
$= E_{ph} = h\nu$

A convenient unit for photon energies $E_{ph} = h\nu$ is the electronvolt, $1 eV = 1.602 \times 10^{-12}$ erg (see also Block 2, Section 2.1.5). It is also common practice to characterize the photon energy by a temperature $T = E_{ph}/k$. The typical thermal energy of particles in a gas with this temperature is the same as the photon energy. Although the concept of a radiation temperature may sound odd it is enormously useful when we have to estimate the radiation expected from a gas or plasma at a certain temperature. You should remember the following rough estimate as a rule of thumb:

$1.602 \times 10^{-12} erg =$ *[energy]* 1 eV corresponds to 10^4 K *[temperature]*

[handwritten left margin:]

⑤ $1 eV = 1.602 \times 10^{-12} erg$

$T = \frac{E_{ph}}{k}$

if $E_{ph} = 1 eV$, $T = \frac{1 eV}{1.381 \times 10^{-16} erg\,K^{-1}}$

$= \frac{1.602 \times 10^{-12} erg}{1.381 \times 10^{-16} erg\,K^{-1}}$

$= 1.160 \times 10^{4} K$

$T \sim 10^4 K$

Activity 2
ss. photon Es + T.
comments p195

Question 5

Verify this rule of thumb by using the conversion from erg to eV and the value of the Boltzmann constant k given in the Appendix. ∎

It is a good idea to get a feeling for photon energies and temperatures in different parts of the electromagnetic spectrum. This is the subject of the first spreadsheet exercise in this block.

Activity 2 (1 hour)

Photon energies and temperature

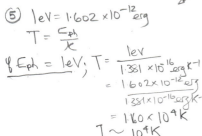

Start up the StarOffice™ software on your computer. If you need help on using spreadsheets, consult Block 1 Activity 3 or Block 2 Activity 3. When you have finished, or if you get seriously stuck, you may want to compare your work with a sample spreadsheet solution which is given in the multimedia resources.

(a) From the frequency axis (horizontal-axis) of the electromagnetic spectrum shown in Figure 9, read off a typical value of the frequency ν for radiation in the radio, infrared, optical, UV, X-ray and γ-ray band. (The X-ray band starts at about 2×10^{16} Hz, the γ-ray band at about 10^{20} Hz). A typical value for visible light would be about 10^{15} Hz. Enter the name of the frequency band in the first column, and your frequency values (in Hz) in the second column of the spreadsheet.

(b) Calculate the corresponding wavelength (in nm) in the third column. To do so you have to click on the first cell of the third column, and enter the equation $\lambda = c/\nu$ to calculate the wavelength in the appropriate units. As $c = 3 \times 10^{17}$ nm s^{-1} the wavelength in units of nm is just $\lambda/nm = 3 \times 10^{17}/(\nu/Hz)$. So you have to enter the string '=3e17/B1' in the first cell of the third column. The 'B1' stands for the first element in column 2, while 3e17 is a conventional notation for 3×10^{17}. Then use the left mouse button to increase the size of the frame that surrounds the cell in the third column that you have just edited such that it contains the whole second column. To do so, click on the lower-right corner of the frame, hold the mouse button down, and drag it to the bottom of the column. When you release the mouse button, it copies the formula into all cells, and the wavelengths for all frequency values should appear.

(c) Repeat the steps above to calculate the energy $E_{ph} = h\nu$ (in eV) in the fourth column.

(d) Then repeat the steps again for the temperature $T = E_{ph}/k$ (in K) in the fifth column.

(e) Express all of these quantities in more appropriate units (for example, if the energy is of order 10^8 eV, then a more appropriate unit would be MeV, giving a value of order 100 MeV). Add three new columns for λ, E_{ph} and T, and enter all values, including the units, by hand.

(f) Finally, for each temperature, find a blackbody curve in Figure 9 that corresponds to a similar temperature. How does the frequency where this curve reaches its maximum relate to the frequency you started with?

You may wish to print a hardcopy of the final spreadsheet and keep it as a 'conversion chart' between photon energies, temperature and frequencies for future reference. (printed off)

Keywords: none

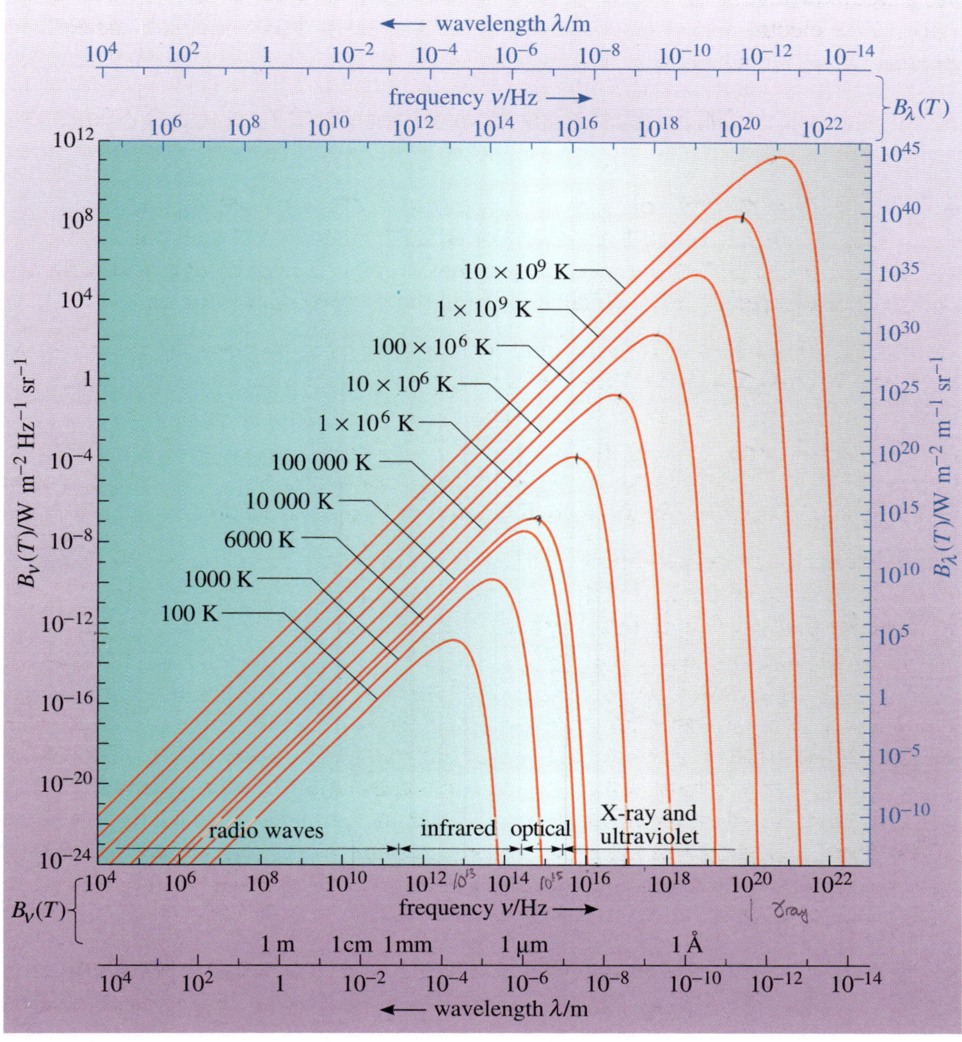

Figure 9 Blackbody electromagnetic spectra.

⑥ Tvirial is the T of the accreted material reached
if its gravitational PE were turned entirely into thermal E.

T_b is the T the source would have if it
were to radiate the same as a BB spectrum

T_b is usually < T_{virial}.

The significance of these Ts they give an
indication of the expected T radiation (the mean
energy of the photons the accreting gas emits)
If the gas is opaque (optically thick) the E
of both the particles & photons is thermalized
(distributed over a wide range of Es)
In this case the $T_{radiative}$ = T_b.
If the gas is transparent (optically thin) then
the photon E distribution is much narrower
$T_{radiation}$ is now the T_{virial}.

Question 6

In the context of mass accretion onto a compact object, explain what is meant by virial and blackbody temperature. Which one is lower? ∎

1.3 Compact stars like it hot

— Compact binaries have a huge diversity *

Algol paradox
— the more evolved star is the least M.

Before we look at the physics of close binaries in more detail you should first appreciate the huge diversity present in the world of compact binaries.

The first star ever recognized as being an interacting binary is **Algol**, an eclipsing binary with an orbital period of 68 hours. It is the prototype of the class of Algol variables. The **light curve** of Algol is shown on page 9 of Collins. Algol is famous for the **Algol paradox** – the more massive component is still in the evolutionary phase of core hydrogen burning, while the less massive companion star has already left the main sequence. This appears to be in conflict with the inescapable fact that more massive stars evolve faster than less massive stars (see Block 2, Section 5.1).

The Algol paradox arises as a consequence of an earlier mass-transfer episode that changed the mass ratio. The evolved companion used to be the more massive star and did indeed evolve faster, but transferred a large fraction of its mass to the other star which therefore is now the more massive component. Mass transfer is still continuing today, and there are signs of a direct impact of the gas stream on the surface of the accreting star. In fact, the gas stream impacts onto the star before it can form a disc, so there is *no* accretion disc of the type we will discuss in detail later in this block. Binaries in such an evolutionary phase are collectively known as Algol binaries (Figure 10).

Algol binaries
— gas stream impacts into star surface
before it can form a disc – ie there is no
accreting disc.

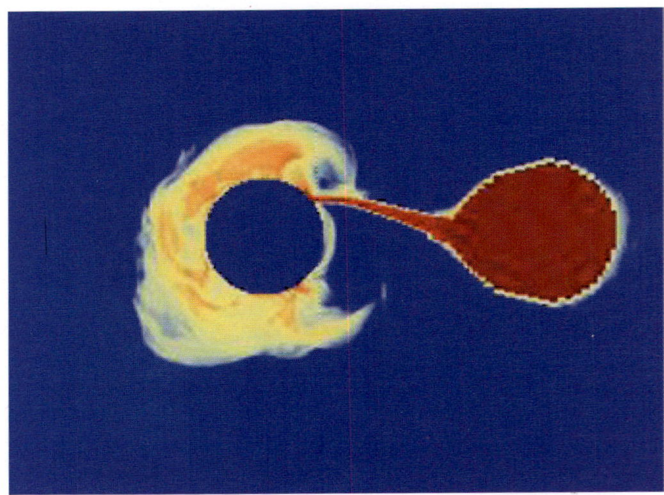

Figure 10 A numerical simulation of mass transfer in an Algol binary. Some of the transferred gas ends up in a tenuous disc-like structure (*ie not* an accretion disc) surrounding the accreting star.

Cataclysmic variables (CVs)
WD accretes mass from low M ms star

many CV show semi regular variation in L
— dwarf nova outbursts, dim & bright periods
lasting a few days & recurring every few weeks
Most likely cause — Instability in accretion disc

1.3.1 Cataclysmic variables

You have already met (in Chapter 1 of FKR) the archetype of compact binaries *with* an accretion disc: in **cataclysmic variables**, or **CVs**, a white dwarf accretes mass from a low-mass main-sequence star. The image on the cover of this Study Guide is an artist's impression of how a CV might look like from a spaceship cruising close to the CV. (You will see other images of compact binaries later in this block, Figures 14, 15, 50 and 51.) Many CVs show semiregular variations of their luminosity, so-called dwarf nova outbursts. The **dwarf novae** alternate between bright states (outbursts) and dim states (quiescence). The outbursts typically last a few days and recur every few weeks. The system SS Cygni is the best studied example.

Observations by amateur astronomers document the sequence of outbursts over a long period of time (Figure 11). We shall talk about the most likely physical cause of dwarf nova outbursts – an instability of the accretion disc – in more detail later in this block.

Figure 11 The long-term light curve of SS Cygni, spanning almost 100 years! (Constructed from observations made by the AAVSO (American Association of Variable Star Observers). Courtesy John Cannizzo)

Some CVs have much more powerful outbursts, so called (classical) **nova** outbursts. As a result of mass transfer, the white dwarf constantly accumulates fresh, hydrogen-rich material on its surface. When the pressure at the bottom of this thin hydrogen-rich layer exceeds a critical limit, hydrogen burning starts under highly degenerate conditions and triggers a thermonuclear runaway. (This is similar to the cause of the helium flash, see Block 2, Section 6.5.) The sudden and violent nuclear burning causes the luminosity to rise by a factor of 1000 or more within hours or days. Figure 12 shows a typical outburst light curve. The affected outer layers of the white dwarf are eventually ejected into space and become visible as a **nova shell**, a dispersing cloud of nuclear-processed gas. The name 'nova' (Latin for 'new', denoting a 'new' star) derives from the fact that some novae in outburst are visible to the naked eye, while the progenitor, an ordinary CV, usually is not (Figure 13). No classical nova has ever been seen to recur. The recurrence time of the underlying outburst phenomenon must therefore be very long, in excess of 1000 years. Likewise, the term 'dwarf nova' has been introduced for obvious reasons, although the physical mechanisms causing classical nova and dwarf nova outbursts are very different from each other.

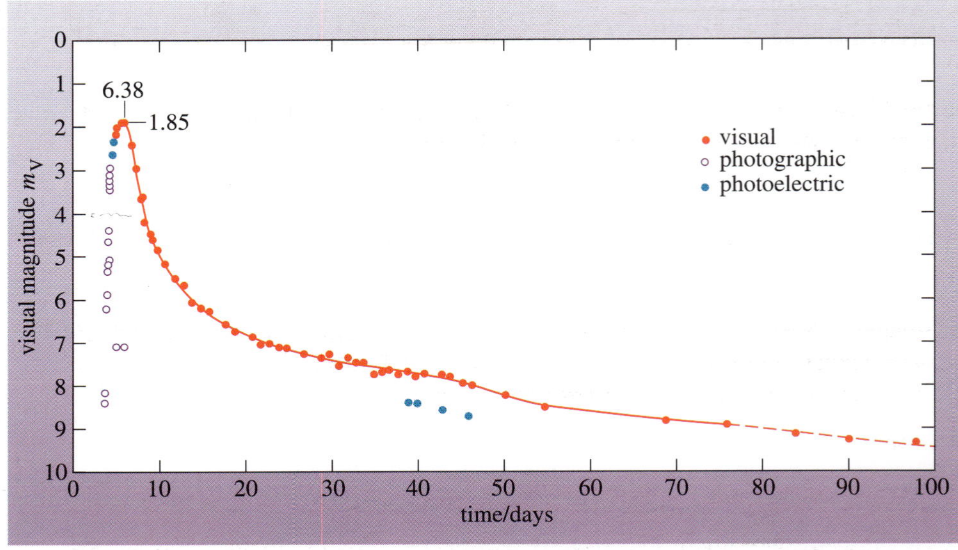

Figure 12 Light curve of V1500 Cygni, a classical nova which went into outburst in 1975.

Activity 3

A nova in Aquilae

From *The Energetic Universe* MM guide, view the animation 'A nova in Aquilae'.

This animation enables you to 'blink' between two images of a star field, with and without nova V1494 Aquilae. This is an animated version of Figure 13. ■

Question 7

From the light curve of SS Cygni in Figure 11, estimate the typical (average) outburst amplitude in magnitudes. By what factor does the visual luminosity rise during the outburst? Compare this with the outburst amplitude of a classical nova, for example, the one shown in Figure 12. ■

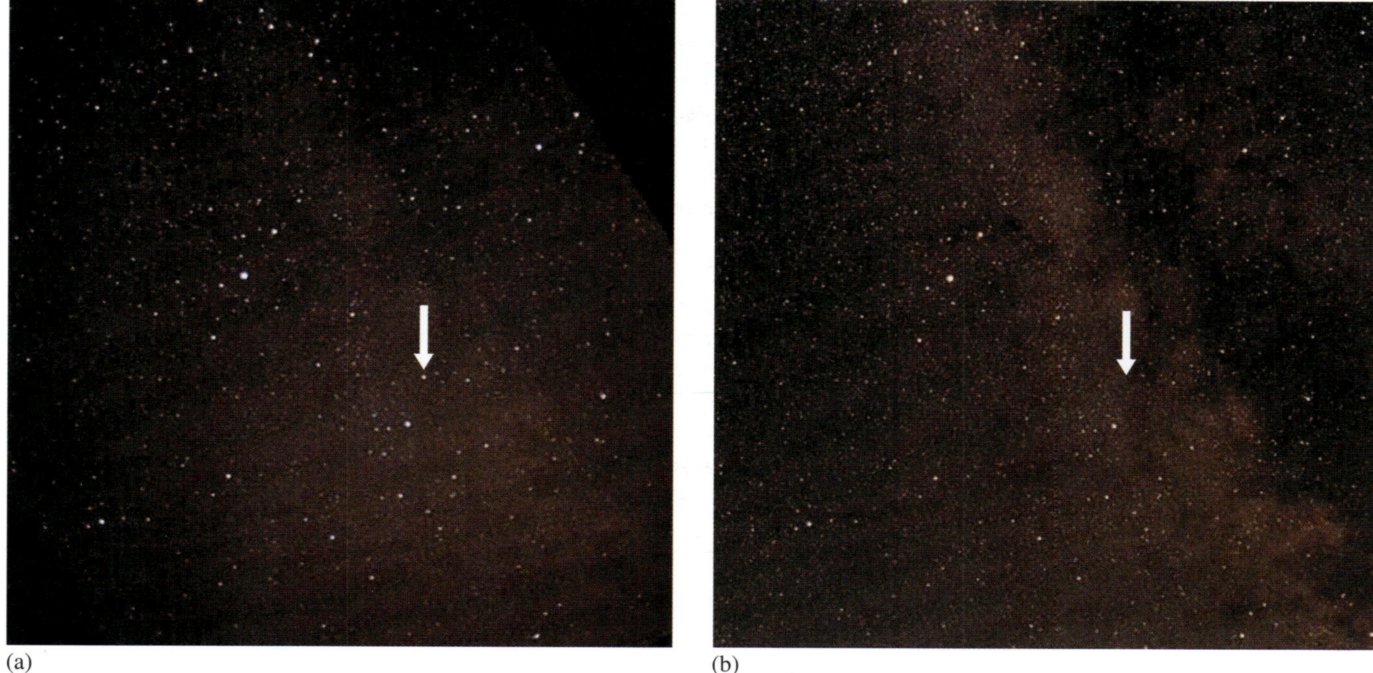

(a) (b)

Figure 13 Star field (constellation Aquilae) (a) with and (b) without nova V1494 Aquilae. You can also 'blink' between these images in Activity 3. (© Till Credner, AlltheSky.com)

Activity 4 (30 minutes)

Nova shells

Find images of nova shells in the Image Archive, and compare them with images of typical planetary nebulae and supernova remnants. (See the 'Comments on activities' if you had difficulty finding the images.)

Keywords: none ■

p.195

Question 8

Describe the morphology of the nova shells you found in the Image Archive. Are they similar to supernova remnants or planetary nebulae? ■

Question 9

From the information given in the caption of the two images of Nova Cygni 1992, calculate the distance to the nova. ■

In the following example we study wind losses from novae.

Example 2

Theoretical calculations of hydrogen-accreting white dwarfs show that a thermonuclear runaway is triggered when the mass of the accumulated hydrogen layer on the white dwarf reaches a value of order $10^{-5}M_\odot$. Compare this with the mass of hydrogen that actually burns into helium during an outburst. Use a typical luminosity ($10^4 L_\odot$) and typical outburst duration (1 yr) for your estimate. What do you conclude from this comparison?

[Handwritten notes in right margin and bottom:]

Comment p 195

Activity 4.
Image Archive
Nova shell images
cpd with PN
 SN remnants

⑧ There are 2 images of nova shells
Nova Cygni 1992 + T Pyxidis.
At low resolution both nebulae appear quite smooth + almost spherical
At higher resolution more structure becomes apparent: T Pyx is v. inhomogeneous (clumpy) + N Cyg develops a pronounced shell structure.

⑧ SN remnants are generally more irregular + larger (actual + angular size)
PN are usually v. symmetrical, but also v. diverse, Some resemble nova shells.

⑨ The nova shell is seen to expand
In the 7 mths between images the mean radius has increased by 0.037". From the v. radial velocity measure of the true expansion velocity is 970 km s⁻¹
The true radial increase of nova shell in 7 mths
$\Delta r = V_{exp} \Delta t = 970 \times 7 \times 30 \text{ day mth}^{-1} \times 86400 \text{ s day}^{-1}$
$= 1.76 \times 10^{15}$ cms.
Distance to nova = D
The angle $\delta = \Delta r/D$ (small angle measure)
$= 0.037"$ (measured)

[Bottom handwritten:]
$D = \dfrac{\Delta r}{\delta} = 9.8 \times 10^{21}$ cms.
Since 3.086×10^{18} cms = 1 pc
$D = 3.2$ Kpc.

23

Solution

To sustain the nova luminosity L for a time Δt by nuclear fusion of hydrogen a certain amount ΔM of hydrogen must burn to helium. As the efficiency of hydrogen burning is roughly 0.7% (see Question 4) we must have

$$L = \frac{0.007 \times \Delta M c^2}{\Delta t}$$

The outburst lasts typically for, say, 1 year. Then the burning mass is

$$\Delta M = \left(\frac{0.007 \times c^2}{L\Delta t}\right)^{-1} = \frac{10^4 \times 3.83 \times 10^{33}\,\mathrm{erg\,s^{-1}} \times 365.25 \times 86\,400\,\mathrm{s}}{0.007 \times (3 \times 10^{10}\,\mathrm{cm\,s^{-1}})^2}$$
$$= 1.9 \times 10^{26}\,\mathrm{g}$$

This is $\Delta M \approx 10^{-7}\mathrm{M}_\odot$, i.e. *much* smaller than the mass accumulated prior to the outburst. The main reason for this discrepancy is that strong wind losses during the burning phase quickly deplete the top hydrogen layer. ■

In Block 2 of this course you have met an even more powerful explosion: supernovae (Block 2, Section 8.2). A supernova of type II is a massive star's death throe, the aftermath of the collapse of its burnt-out core, while a supernova of type Ia involves a compact white dwarf binary. A type Ia supernova occurs when a white dwarf is pushed over the Chandrasekhar limit, triggering a fatal collapse. Type Ia supernovae are thought to be **standard candles**, allowing one to determine the distance to remote galaxies, and hence to measure the size and expansion of the Universe.

1.3.2 X-ray binaries

Up to now we have met only systems where the compact star is a white dwarf. But what happens if the accreting star is an even more compact neutron star or a black hole? Since the accretion efficiency $\eta = GM/Rc^2$ is about 1000 higher for neutron stars and black holes than for white dwarfs, the emitted photons have much higher energies. Consequently, neutron star and black hole binaries appear as X-ray sources and are collectively known as X-ray binaries (Figures 14 and 15). The mass M_2 of the compact object's companion star can be used to separate X-ray binaries into three main groups with distinct properties which we will discuss in the next section. In **low-mass X-ray binaries** the companion star is a low-mass star ($M_2 \lesssim 2\mathrm{M}_\odot$), while in **high-mass X-ray binaries** the companion star is a massive star ($M_2 \gtrsim 10\mathrm{M}_\odot$). The companion star mass in **intermediate-mass X-ray binaries** is – you guessed it – roughly in between these limits.

Many X-ray binaries display variations in their X-ray flux on quite a range of timescales. These contribute to the large number of variable 'stars' that appear and disappear on the X-ray sky, as you will see in the next activity.

Activity 5 **(15 minutes)**

The X-ray sky

From *The Energetic Universe* MM guide, view the movie entitled 'The X-ray sky' which shows the X-ray sky as observed by the XTE X-ray satellite.

The animation sequence displays the entire sky (above) and the central part of the Milky Way Galaxy (below). Each 'frame' of the movie displays the X-ray sky at twelve-hour intervals, at a rate of 4 days per second. A circle is plotted for each X-ray source observed by the All-Sky Monitor aboard the XTE satellite. The area of circle is proportional to the intensity of the source. The colour of the circle is indicative of the temperature of the source: blue indicates higher temperatures and red cooler.

Keywords: none ■

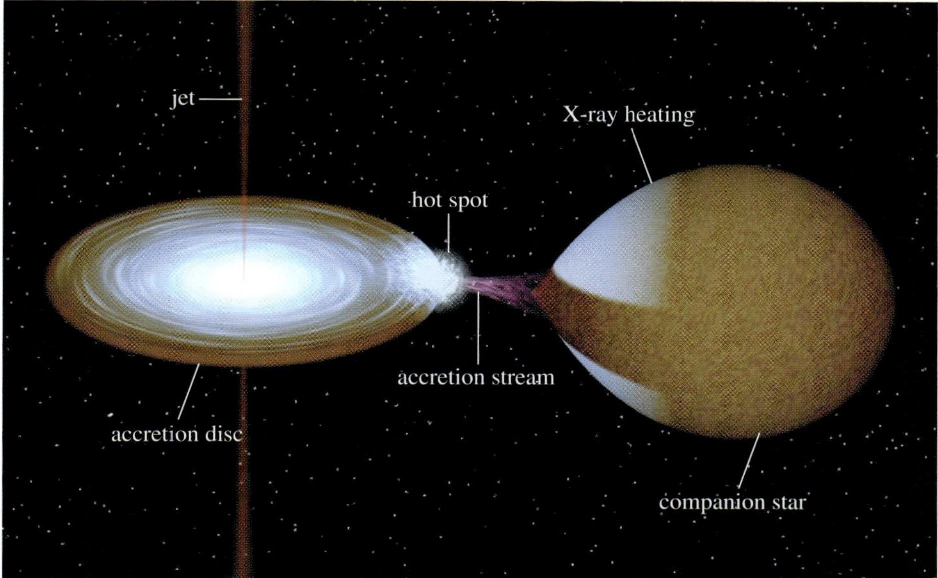

Figure 14 Impressions of X-ray binaries: the low-mass X-ray binary V1033 Scorpii (also known as GRO J1655-40), a superluminal jet source. The black hole accretes matter from a Roche-lobe filling low-mass or intermediate-mass companion star. The orbital period is 2.6 days. (Courtesy of Rob Hynes)

Figure 15 Impressions of X-ray binaries: the high-mass X-ray binary Cygnus X-1. The black hole accretes from the wind of the massive companion star. The orbital period is 5.6 days. (Courtesy of Rob Hynes)

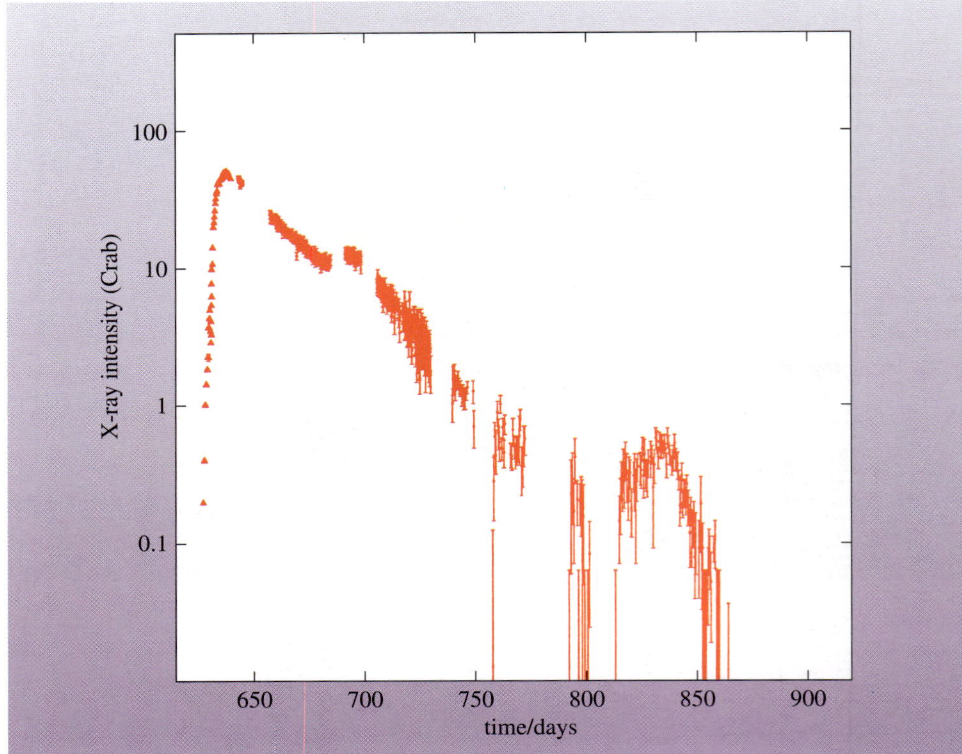

Figure 16 X-ray light curve of the black hole X-ray binary A0620-00 (Nova Monocerotis 1975 also known as V616 Monocerotis), a soft X-ray transient. The X-ray intensity is given in multiples of the Crab nebula intensity. The time axis gives days since Julian Date JD244 2 000.

A particularly conspicuous group of X-ray binaries, the **soft X-ray transients**, will be subject to closer scrutiny later in this block. Confusingly, they are also called X-ray novae, as their X-ray light curves resemble the optical light curve of classical novae, and the X-ray outburst is usually also accompanied by an optical outburst. During outburst the X-ray flux often increases very quickly by a factor of 1000 or more (Figure 16). This is usually followed by a slow decline over several months into quiescence, where systems are in many cases below or close to the detection limit of present-day X-ray telescopes. Most of the soft X-ray transient systems have been observed to go into outburst only once, while in some of them outbursts recur quasi-periodically every few years or decades. Soft X-ray transients can have either a neutron star or black hole accretor. The outburst is thought to arise from an instability in the accretion disc. Figure 17 summarizes the black hole soft X-ray transients known at the time of writing.

A small number of X-ray binaries have been observed to eject matter either episodically or continuously in the form of a pair of tightly focused **jets** in opposite directions (see Figure 14). The jets are thought to form in the vicinity of the compact star, and material in the jet initially propagates along an axis of symmetry in the system, the rotational or magnetic field axis of the compact object. Most amazingly, in some of the X-ray binaries with jets the speed in the jet seems to exceed the speed of light, in clear conflict with Einstein's theory of relativity! But relax, the actual speed of the material is thought to be close to (but less than) c, while the apparent superluminal motion is a consequence of a projection effect and the constancy of the speed of light. Superluminal jets are also seen in active galactic nuclei, and will be discussed in this context in Section 3.4.5 of Block 4. The Galactic superluminal jet sources are sometimes called 'microquasars'.

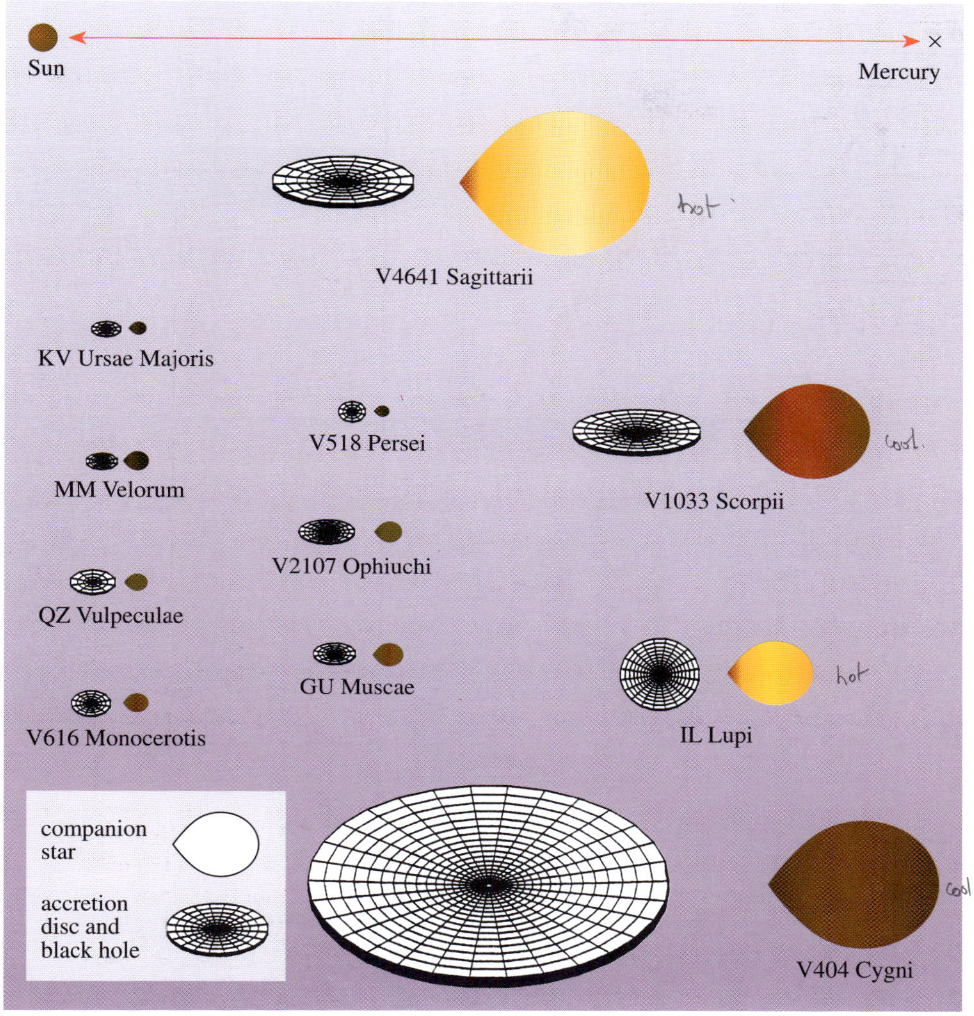

Figure 17 Soft X-ray transient systems with a black hole accretor, on a scale based on the distance between the Sun and Mercury, indicated at the top of the figure. The colour of the companion (donor) star indicates its surface temperature: dark red is cool, bright yellow is hot. (Courtesy of Jerry Orosz)

Activity 6 **(20 minutes)**

Superluminal jet sources

From *The Energetic Universe* MM guide, watch the animation sequence 'Superluminal jet sources' and read the caption.

The sequence shows images in the radio waveband of the jet ejections of the most famous microquasar GRS 1915+105, a mysterious black hole binary on the far side of the Galaxy (in the constellation Aquilae).

Keywords: **superluminal jet source** ■

Activity 6
The Energetic Universe MM guide
Superluminal jet sources
animation sequence

Superluminal jet source
an X-ray binary or active galaxy displaying a jet that appears to expand with speeds > c.
Superluminal motion
the apparent faster than light motion exhibited by some superluminal jet sources
The obs motion is due to material emitted at a small angle to line of sight at a speed < c

jet used in the description of radio source structure with 2 related meanings.
① a long narrow linear feature emanating from the AGN seen in many radio galaxies + quasars, most often via radio synchrotron emission, but also at optical x X-ray λs
② the underlying flow of energetic particles from the nucleus supplying E to radio lobes, present in all radio sources whether or not jets are observed

Question 10

From the information in the caption provided with the animation sequence (Activity 6), answer the following questions: (a) At what wavelength were these observations made? (b) How many days do these observations cover? (c) In this time, what angular distance do the blobs travel on the sky? (d) The distance to the binary is 12.5 kpc. Calculate the apparent expansion speed in the jet, assuming that the material in the jet moves at right angles to the observer. ■

radio band $\lambda = 3.6$ cms
outburst on March 18, 1994, obs over April 16, 1994, last obs 28 days after outburst
angular distance travelled by each $\delta = 0.5''$

D = 12.5 Kpc, Implied linear expansion $\Delta l = D \tan\delta$, Implied velocity, $v = \Delta l / 28$ days

$\frac{D\tan\delta}{\Delta t} = v = \frac{\tan(0.5'') \times 12.5 \times 10^3 \times 3.086 \times 10^{18} \text{ cms}}{28 \times 86400 \text{ s}} = 1.29 \times c$ in a vacuum.

Variability of the X-ray flux on much shorter timescales is observed in some neutron star X-ray binaries. They display outbursts lasting seconds to minutes that recur every few minutes (Figure 18, and FKR Figure 6.20). These so-called **X-ray bursts** (of type I) represent the thermonuclear burning of the accreted hydrogen-rich material – they are the analogue of classical novae in CVs. The shorter timescale and more energetic radiation arise because the burning is triggered much deeper in the potential well of the accreting star. If type I X-ray bursts are observed from an X-ray binary, they are taken as unambiguous proof of the presence of a neutron star. A black hole does not have a hard stellar surface, so any accreted material silently crosses the **event horizon** and disappears down the hole.

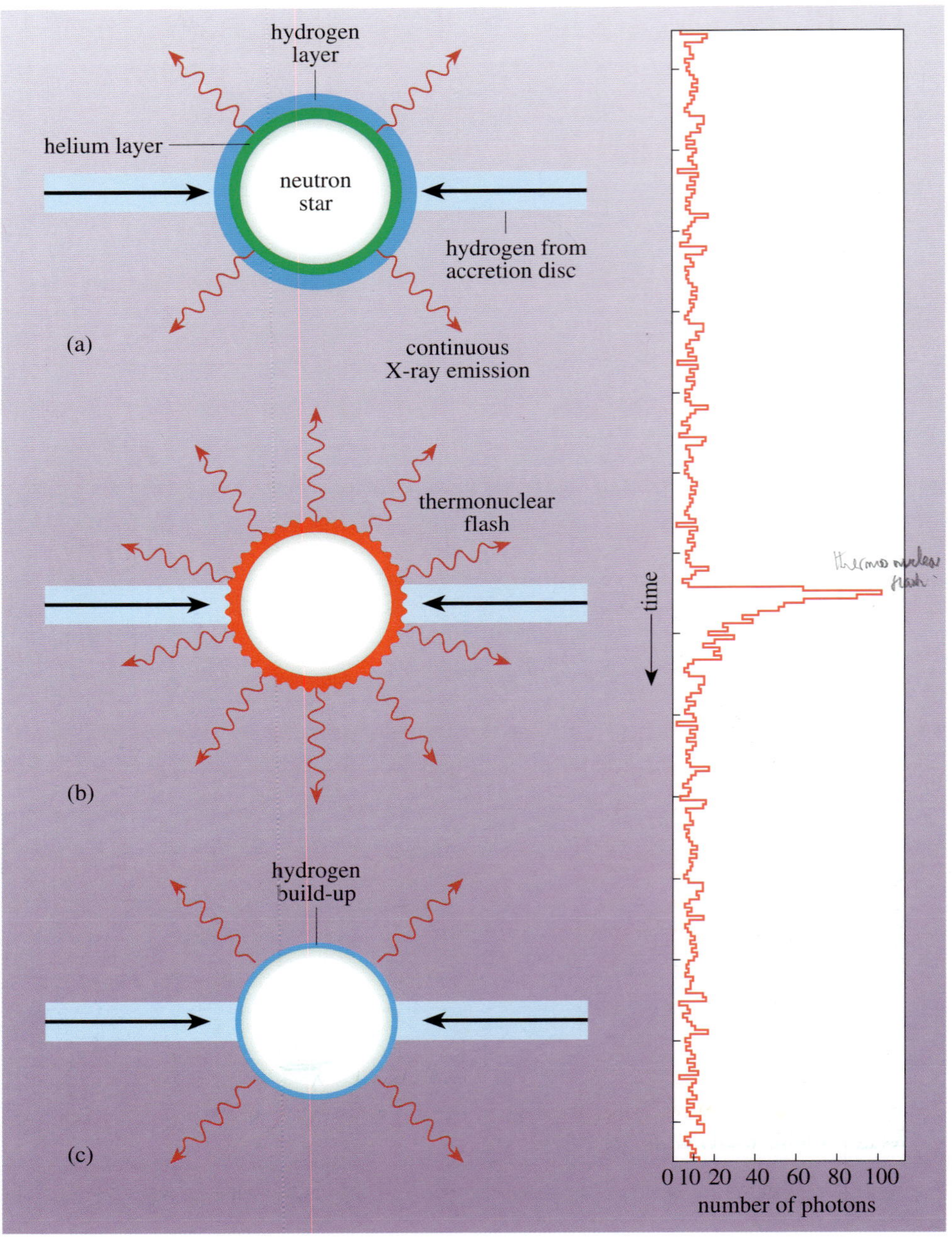

Figure 18 The origin of X-ray bursts. The right-hand panel shows a typical X-ray light curve.

Question 11

Explain the difference between a dwarf nova, a classical nova and an X-ray nova, and the difference between a type I X-ray burst and type Ia supernova. ■

NOTE Are you confused by this plethora of names and designations, the variety of novae and the like? Perhaps the schematic overview in Figure 19 or summary in Table 1 (overleaf) will help to make it clearer? If not there is no need to worry, further on in the block you will see what all these phenomena have in common – the physics of accretion in interacting binary stars.

[Handwritten margin notes:]

dwarf nova are CVs – systems with WD accretors. that alternate semi regularly between outbursts + quiet on timescale of ~ wks. Outburst due to disc instability

Classical nova occur in CVs. Constitute explosive burn of H+He on WD surface, Much larger outburst amplitude + none have gone into outburst more than once. (theoretically outbursts could be $10^4 - 10^5$ yrs)

X-ray binary system if NS + BH is the accreting star. X-ray novae analogue of dwarf novae. Outburst due to instabilities in disc.

Type I X-ray bursts are the NS analogue of classical novae (X-ray burst occur frequently)

Type Ia Supernova is a disruption of a WD driven over the Chandrasekhar mass – could occur when mass accretion. It can only occur once.

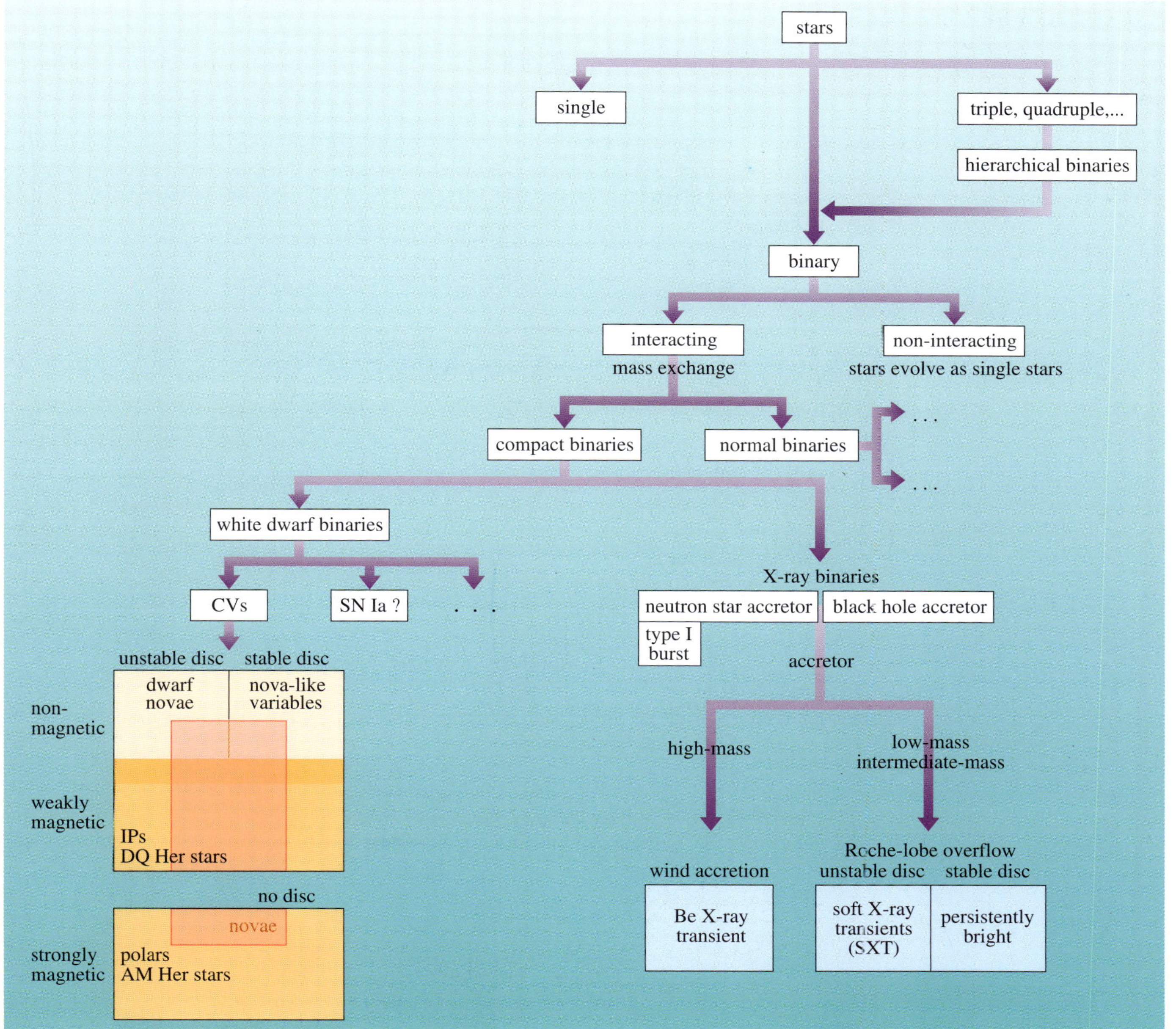

Figure 19 A schematic summary of the variety of binary and single stars. Some of the CV subclasses will be introduced in Section 6.

Table 1 Nomenclature and outburst recurrence times for compact binaries.

Outburst mechanism	System (compact star)	
	CV (white dwarf)	X-ray binary (neutron star)
accretion disc instability	dwarf nova $\Delta t \approx$ days to weeks	X-ray nova $\Delta t \approx 10$ yr
thermonuclear runaway	classical nova $\Delta t \gtrsim 10^4$ yr	type I X-ray burst $\Delta t \approx$ hours

A word of warning is called for: our discussion of the phenomenology of X-ray variability is far from complete. Our selection is based on what you will encounter in the readings of FKR and in other activities of this block. Should you browse the research literature on X-ray binaries (or other textbooks), you are likely to meet other phenomena on yet shorter timescales. In particular you may find reference to the pulsed X-ray emission from **X-ray pulsars** due to the rotation of the accreting neutron star, and the less regular so-called quasiperiodic oscillations (commonly abbreviated as QPOs). As these are not subject of the course you do not have to know what these are, but you are welcome to read the corresponding entries in Collins.

Before we come to the summary of the first main section of this block, we introduce an 'activity' that we shall revisit at the end of each section.

Self-study suggestion *The name game: part 1*

We suggest you skim-read Section 1 again, and as you do so make a list of all symbols that denote *new* variables and *new* quantities, and note down what they mean. Your list should at least include L_{acc}, $\dot{M}$, R_*, L_{edd} and η.

Once you have compiled your list and explained the meaning of all the entries, compare it to the one given in Appendix A5 at the end of the Study Guide.

We will repeat this task at the end of each section so you will be able to use this list as a resource while you study the block. If you feel you are getting lost and struggle to remember the meaning of a certain symbol – you know where to look for help.

1.4 Summary of Section 1

1 Binary stars or hierarchical binaries are the only stable configurations of gravitationally bound several-body stellar systems.

2 Estimates for the binary fraction range from 20% to more than 90%.

3 A stellar companion effectively constitutes a measuring device to determine the masses of binary components, in eclipsing systems sometimes even the radii of the stars.

4 Mass transfer from one component to the other ensues when the orbital separation is small enough. If the mass-gaining star is compact (its mass to radius ratio is very small) the accreting material becomes very hot. Temperatures in excess of 100 000 K for white dwarfs and tens of millions of kelvin for neutron stars and black holes are the norm. Neutron star and black hole binaries appear as X-ray sources and are collectively known as X-ray binaries.

5 A plasma is a conducting fluid whose properties are determined by the presence of ions and electrons in the fluid.

6 Stellar plasma leaves the donor star from the vicinity of the inner Lagrangian (L_1) point. Because of the conservation of angular momentum the plasma stream swings into an orbit around the compact, accreting star. An accretion disc forms.

7 Accretion discs are like machines that extract gravitational potential energy and angular momentum from plasma.

8 The accretion luminosity is

$$L_{\rm acc} = \frac{GM\dot{M}}{R_*}$$

9 The maximum luminosity of a gravitationally bound spherical object is the Eddington luminosity

$$L_{\rm edd} = \frac{4\pi GMm_{\rm p}c}{\sigma_{\rm T}} \approx 1.3\times10^{38}\left(\frac{M}{\rm M_\odot}\right){\rm erg\,s^{-1}}$$

10 The accretion efficiency η is defined by

$$L_{\rm acc} = \eta\dot{M}c^2$$

11 For neutron stars and black hole accretors the accretion efficiency is about 10–20%. Accretion is the most efficient energy source in the Universe.

12 In Algol binaries the more massive component is still in the phase of core hydrogen burning, while the less massive companion star has already left the main sequence. This is the result of a previous phase of mass transfer.

13 In cataclysmic variables (CVs), a white dwarf accretes mass from a low-mass main-sequence star. Dwarf novae, a subtype of CVs, alternate between outbursts (days) and quiescence (weeks). A classical nova outburst constitutes a thermonuclear runaway in previously accreted top layers on the surface of a white dwarf in a CV. These layers are eventually ejected into space and become visible as a nova shell.

14 Soft X-ray transients (also known as X-ray novae) are X-ray binaries that undergo strong outbursts, with an increase of their X-ray flux by a factor of 1000 or more. This is usually followed by a slow decline over several months into quiescence. Most of the soft X-ray transient systems have been observed to go into outburst only once. Soft X-ray transient outbursts are analogous to dwarf nova outbursts in CVs.

15 Some X-ray binaries eject matter in narrowly focused jets. In superluminal jet sources the speed in the jet appears to exceed the speed of light, but this is a projection effect.

16 Some neutron star X-ray binaries display outbursts lasting seconds to minutes that recur every few minutes to hours. These X-ray bursts (of type I) arise from the thermonuclear burning of the accreted material. Type I X-ray bursts are analogous to classical nova outbursts in CVs.

2 STARS PLACED IN A CAGE

Clearly, when two stars orbit each other in close proximity, neither of them can get arbitrarily large without feeling the restrictive presence of the second star. If one star becomes too large, the gravitational pull on its outer layers from the second star will become bigger than the pull towards its own centre of mass. Then mass is lost from one star and transferred to the other star. (See the box on 'Binary stars'.)

BINARY STARS

The binary component losing mass to the other component is called the mass **donor**, while the component on the receiving end is called the mass **accretor**. Quite often the accretor is also referred to as the primary star, or just the **primary**, while the mass donor is the secondary star, or simply the **secondary**. This is because in many cases (but not in all cases!) the accretor is more massive than the donor. We shall adopt this nomenclature in this Study Guide, and so does FKR. Quantities carrying the index '1' usually refer to the primary, while those with index '2' refer to the secondary. The **mass ratio**

$$q = \frac{M_2}{M_1} \tag{5}$$

is therefore the ratio of donor mass to accretor mass, and usually (but not always) less than unity. But beware! This is not a generally accepted convention, and other books or journal papers may use a different one. So if you read about binary stars and encounter terms like 'primary star' and 'mass ratio', be sure to check if star 1 or star 2 is the mass donor, and if the mass ratio is defined as M_2/M_1 as here, or as M_1/M_2. (Even professional astronomers are known to have fallen into this trap.)

This situation is illustrated in Figure 20 which shows the effective potential in the binary star along the axis through the centres of the two stars. Note that this axis is not at rest, but is rotating with the binary system, so centrifugal forces as well as gravitational forces are present. The significance of the effective potential Φ_R is that at each point x along the axis the slope $\partial\Phi_R/\partial x$ (gradient) of Φ_R indicates the direction and magnitude of the force a small test body placed at this point in the binary system would feel. You can see the effect of the centrifugal repulsion at large distances from the binary's centre of mass (Φ_R falls off at large x), and the two deep valleys caused by the gravitational attraction of the corresponding star in the respective valley. A star can fill these valleys only up to the 'mountain pass' in between. If the star attempts to grow further, mass flows over into the neighbouring valley. In a phase with a continuous flow of mass the donor star will fill the maximum volume available to it, its **Roche lobe**. The mass flows to the less extended accretor which resides well inside its own lobe. This mode of mass transfer is therefore known as **Roche-lobe overflow**, and the binary is said to be **semidetached**. Examples are cataclysmic variables and low-mass X-ray binaries.

deep valleys are caused by gravitational attraction of the corresponding star

Φ_R falls off at large x from the binary's centre of mass? due to centrifugal repulsion

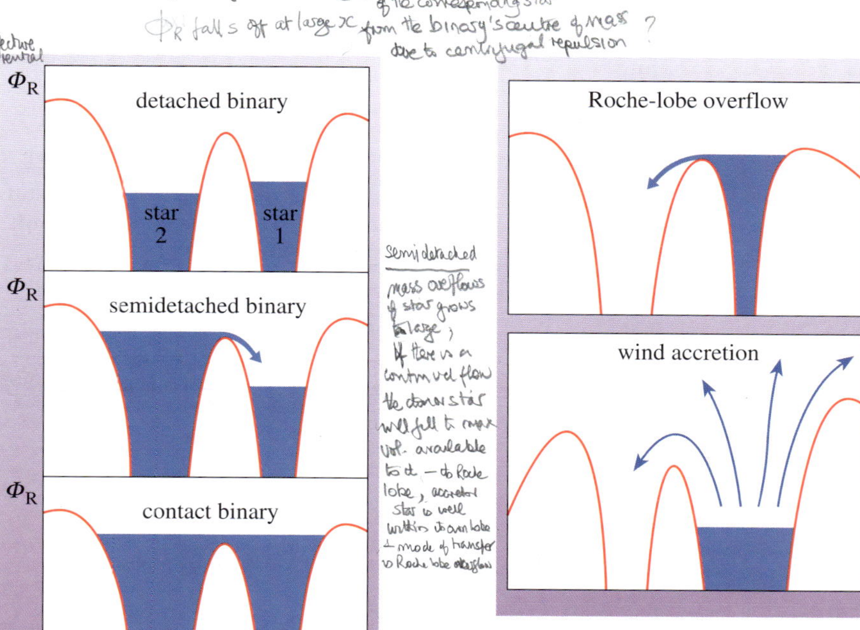

Figure 20 Schematic view of the potential wells of a detached, semidetached and contact binary.

axis rotating with system, (F_G + Fcentrifugal forces present)

semidetached
mass overflows if star grows too large; if there is a continued flow the donor star will fall to max vol. available to it — the Roche lobe, accretor star is well within its own lobe → mode of transfer is Roche lobe overflow

Figure 21 Schematic view of Roche-lobe overflow and wind-fed accretion.

Roche-lobe overflow starts either when one of the stars attempts to grow beyond its lobe, or because the lobe closes in on the star. The former can occur simply as a result of the star's nuclear evolution; as you have seen in Block 2, the radius of a star increases during its life. The latter can occur if the orbit shrinks by losing orbital angular momentum. We shall come back to both possibilities in Section 2.2.

Before we examine Figure 20 and the Roche lobe more quantitatively we note in passing that there is yet another way to establish mass transfer, as indicated in the lower panel of Figure 21. Massive stars and giant stars display rather strong stellar winds (Block 2). The accretor can capture a fraction of the matter the other star loses in its wind. Hence mass is transferred even though the mass-losing star resides well inside its Roche lobe. This mode of mass transfer is called **wind accretion**. Most of the mass in the wind is lost from the binary, though. This is the most common form of accretion in high-mass X-ray binaries.

examples

The images in Figures 14 and 15 have already given you an idea of how binaries undergoing Roche-lobe overflow and wind accretion might look. A variation of wind accretion occurs in Be/X-ray binaries, a subgroup of high-mass X-ray binaries. In these systems a neutron star orbits a massive **Be star** on an elliptical orbit. At **periastron**, i.e. at closest approach on the elliptical orbit, the neutron star accretes material from the Be star's wind (Figure 22), and periodically appears as a (hard) X-ray transient.

example

Be stars
Spectral type B
They are irregularly variable + show strong emission lines in their spectra. The emission is often polarized. + its IR emission is often stronger than other B stars. These unusual properties are a result of a strong stellar wind which often forms a circumstellar disc. Some are massive YO similar to T Tauri

periastron (accretes material from Be star wind - sometimes appears as X-ray transient)

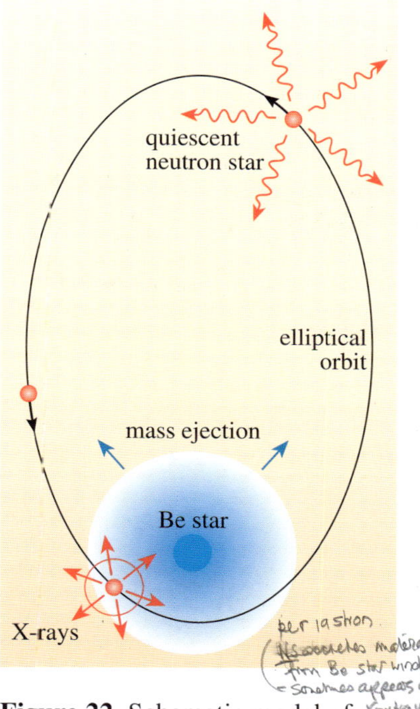

Figure 22 Schematic model of how accretion might proceed in an eccentric Be/X-ray transient. The rapid rotation of the Be star has created a circumstellar disc which, if the orbit is sufficiently eccentric, the neutron star can pass through. Then accretion may take place, and an X-ray outburst starts.

Activity 7

Interacting binary systems

To conclude this introductory section, read Sections 4.1–4.2 of FKR.

Keywords: none ■

2 main reasons why many binaries transfer matter at some stage of their evolution lifetime
① one star may increase R or the binary separation shrinks to the point where the gravitational pull of companion can remove outer layers (Roche lobe overflow)
② one star in an evolutionary phase may eject much of its mass as stellar wind, some of this will accrete onto companion by gravitational pull

Activity 7
FKR Sections 4.1-4.2
accretion as a power source was first recognised in binary systems (especially X-ray binaries). Majority of all stars are in binaries so at some stage in evolution has undergone mass transfer. Angular momentum is important in accretion - the transferred material cannot land onto if it has lost most of angular momentum - this leads to accretion disc formation - these are efficient machines for extracting gravitational PE + converting it to radiation - central engine in AGN

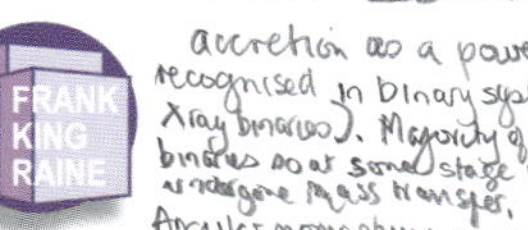

[Handwritten margin notes: Pseudo forces/fictitious forces – forces with no basis in physical reality, but which can be used to account for the motion of bodies observed from non inertial frames of reference. By intro these forces the bodies can be made to conform to Newton's Laws even though these laws do not apply in non inertial frames (eg. centrifugal force)

Inertial frame of reference – a frame in which any particle that is not acted on by an unbalanced forces moves with constant speed along a straight line. (Newton's I law applies). Any frame of reference that moves with constant velocity relative to an inertial frame of ref is also an inertial frame of ref]

2.1 A hole in the balloon: the Roche model

We shall now consider the physical foundation of the schematic Figure 20 in greater detail. Roche-lobe overflow is most conveniently discussed in a frame of reference that co-rotates with the binary. This frame rotates about the rotational axis of the orbital motion, i.e. an axis through the binary's centre of mass and perpendicular to the orbital plane, with the same angular speed ω as the binary,

$$\text{angular speed} \cdot \omega = \frac{2\pi}{P_{\text{orb}}}$$

[Handwritten: $\frac{2\pi}{\omega} = P_{\text{orb}}$, $\frac{2\pi}{P_{\text{orb}}} = \omega$; rotational axis, orbital plane]

where P_{orb} is the orbital period. The co-rotating frame does *not* constitute an **inertial frame**. **Pseudo-forces** appear as a result of the rotational motion. To see why, recall Newton's laws of motion. Any acceleration of a body in an inertial frame results from the action of a force on this body. In contrast, in a rotating frame a body may accelerate relative to the observer simply because the observer himself or herself is fixed to the rotating frame, while the body is not. The force causing this acceleration does not exist in the inertial frame, hence is called a pseudo-force. Nonetheless, for an observer in the rotational frame it can be very real.

In particular, there are two pseudo-forces in a rotating frame, the centrifugal force and the Coriolis force. You are certainly familiar with the **centrifugal force**, which you experience for example in a car that follows a sharp bend of the road at high speed. As the driver you are an observer fixed to the rotating frame, the car. The rotational axis is vertical and passes through the geometric centre of the bend. You are at rest in the driver's seat, but to stay there you need your muscle strength to balance the centrifugal force pushing you radially outwards, away from the centre of the bend. For an observer in the inertial frame – someone standing on the pavement – you are of course not at rest, but travelling on a circular path. The pedestrian concludes that there is a **centripetal force** acting on you, which is pulling you off the straight line (Figure 23). This force is mediated by the friction of the tyres on the road, the structural rigidity of the car, and your muscle strength. In fact, the centripetal force has the same magnitude as the centrifugal force, but the opposite direction. The magnitude of the centrifugal force on a body with mass m and distance r from the rotational axis is

$$F_{\text{c}} = m\frac{v^2}{r} = m\omega^2 r \qquad v = \omega r \tag{6}$$

Here ω is the angular speed of the rotating frame, and $v = \omega r$ the magnitude of the instantaneous velocity of a point fixed to the rotating frame, with respect to the non-rotating inertial frame.

The second pseudo-force in the rotating frame, the **Coriolis force** acts *only* on bodies that are moving in this frame. The Coriolis force is always perpendicular to the direction of motion, and also perpendicular to the rotation axis. It is easy to see why such a force in addition to the centrifugal force must exist. A body at rest in an inertial frame would appear to move in a circle around the rotational axis in the rotating frame (Figure 24b). Hence the observer in the rotating frame concludes that there is a force at work which not only overcomes the outward centrifugal force, but also provides the inward centripetal force necessary to maintain the circular motion.

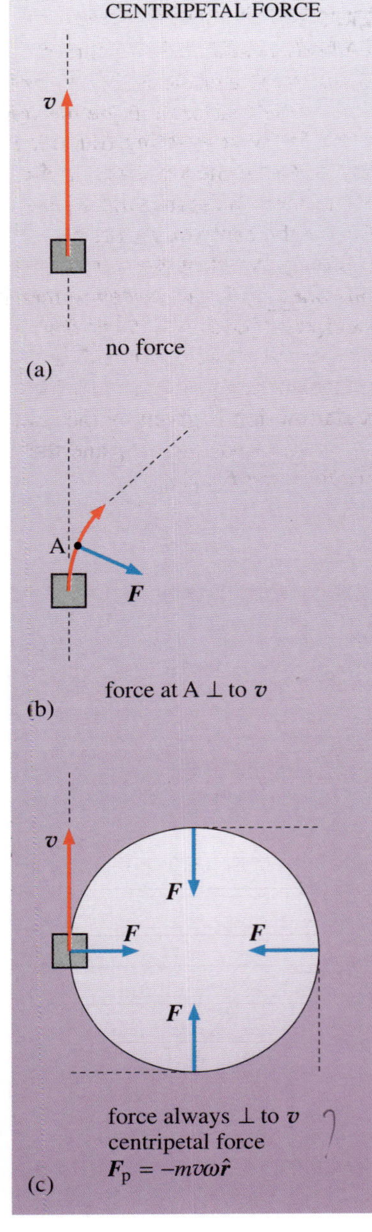

Figure 23 The centripetal force. The observer is in a non-rotating frame of reference (inertial frame).

CENTRIPETAL FORCE

(a) no force

(b) force at A ⊥ to v

(c) force always ⊥ to v
centripetal force
$F_p = -mv\omega\hat{r}$

Pseudo-forces are often called fictitious forces (as in Block 1).

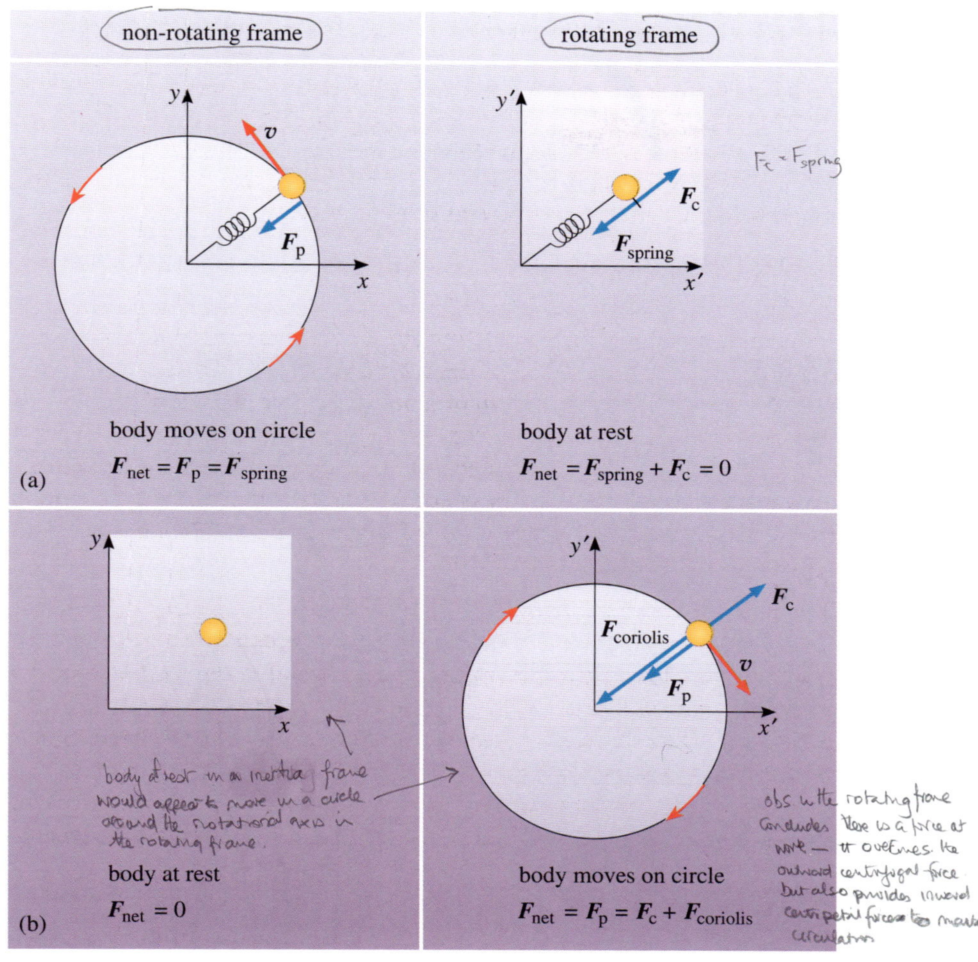

Figure 24 The Coriolis force. (a) A body is attached to a spring and moves on a circle with constant angular speed (as seen in the inertial frame). In the co-rotating frame the body is at rest; the net force on the body is zero: the spring force just balances the centrifugal force. (b) A body at rest in the inertial (non-rotating) frame is seen to move on a circle with constant angular speed in the rotating frame. The necessary centripetal force $\boldsymbol{F}_p$ for circular motion is given by the sum of the centrifugal force $\boldsymbol{F}_c$ and the Coriolis force $\boldsymbol{F}_{coriolis}$.

Example 3

Use the example of a body at rest in the inertial frame to show that the Coriolis force has the magnitude $2m\omega v$. Consult Figure 24.

Solution

The body of mass m is at rest in the non-rotating frame. Its distance from the rotational axis is r. In the frame rotating with angular speed ω the same body appears to move on a circle with radius r and speed $v = r\omega$.

The observer in the rotating frame concludes that there is a net force, the centripetal force $\boldsymbol{F}_p$, of magnitude $m\omega^2 r = m\omega v$ acting on the body. This force points towards the centre of the circle. The observer knows that there are two pseudo-forces acting on the body, the centrifugal force $\boldsymbol{F}_c$ of magnitude $m\omega^2 r = m\omega v$, pointing away from the centre, and the Coriolis force $\boldsymbol{F}_{coriolis}$. From the vector sum

$$\boldsymbol{F}_{net} = \boldsymbol{F}_p = \boldsymbol{F}_c + \boldsymbol{F}_{coriolis}$$

(see Figure 24b) we have

$$\boldsymbol{F}_{coriolis} = \boldsymbol{F}_p - \boldsymbol{F}_c$$

As $\boldsymbol{F}_p$ and $\boldsymbol{F}_c$ point in opposite directions it is clear that the magnitude of the Coriolis force is just the sum of the magnitudes of $\boldsymbol{F}_p$ and $\boldsymbol{F}_c$, i.e. $2m\omega v$, as required.

(The expression for the Coriolis force is slightly more complicated if the velocity is not perpendicular to the rotation axis; it involves the vector product

$$\boldsymbol{F}_{\text{coriolis}} = -2m\boldsymbol{\omega} \times \boldsymbol{v}$$

(see Section 1.12.6 of Block 1), so that only the velocity component perpendicular to the axis enters the calculation.) ■

We do in fact live in a rotating frame of reference ourselves: on Earth. The Coriolis force is responsible for the motion of clouds around high- or low-pressure weather systems, as seen in satellite images of the Earth (Figure 25).

Figure 25 Earth as seen from Space, with a cyclonic depression.

Question 12

Calculate the ratio between centrifugal and gravitational force at the Earth's equator, at the North Pole, and at the South Pole. The Earth's radius is 6.4×10^6 m, the acceleration due to gravity at the Earth's surface is $9.81\ \text{m s}^{-2}$. ■

The magnitude and direction of the centrifugal force depends only on the position in the rotating frame. It can therefore be expressed as the gradient of a potential, such that $-\boldsymbol{\nabla}V = \boldsymbol{F}_{\text{c}}$ (revise Section 3.9 in Block 1 if you are unsure about the meaning of 'gradient'). In contrast, the Coriolis force depends on position *and* velocity, and cannot be derived from a potential. The most important thing to remember about the Coriolis force is that it vanishes if $v = 0$ in the rotating frame!

Inertial frame of reference – in a frame of ref in which any particle that is not acted on by an unbalanced force will move with constant speed in a straight line (Newton's 1st law applies in this frame of ref). Any frame of reference that moves with constant velocity relative to an inertial frame is also an inertial frame of ref.

Coriolis force – a fictitious force used to account for certain aspects of motion of bodies observed from a rotating frame of ref. Effects of the force, it causes bodies moving towards/away from rotation axis to be deflected at right angles to their direction of motion. Introduction of this fictitious force allows bodies to conform to Newton's Law predictions.

Centripetal force – the unbalanced force required to keep a body in uniform circular motion about a fixed point. When a particle of mass m moves around a circle of radius r with uniform angular speed ω (uniform speed ($v = r\omega$)) its acceleration is directed to centre + has magnitude $r\omega^2 = v^2/r$. From Newton 2nd Law it must be acted on by an unbalanced force towards centre with mag. $F = mr\omega^2 = mv^2/r$

Centrifugal force (fictitious force) may be used to account for certain aspects of the motion of bodies observed from a rotating frame of reference. Effect of centrifugal force is to cause bodies to accelerate rapidly outwards from rotational axis. By introducing such fictitious forces the motion of bodies may be made to conform with Newton's Law predictions.

Centrifugal force depends only on position in a rotating frame – expressed as the gradient of a potential ∇V

Mag + direction of centrifugal force depends only on the position in the rotating frame

$\nabla = \left(\frac{\partial}{\partial x}, \frac{\partial}{\partial y}, \frac{\partial}{\partial z}\right)$ grad operator

gradient of a scalar field $T(x,y,z)$ = ∇T is a vector field

$\nabla T = \left(\frac{\partial T}{\partial x}, \frac{\partial T}{\partial y}, \frac{\partial T}{\partial z}\right)$ are the components of ∇T

(12) Centrifugal force on a body $F_c = m\omega^2 r$

Gravitational force $F_{GR} = mg$

where $\omega = \frac{2\pi}{86400\text{ s}} = \frac{2\pi}{P_{orb}}$

$r = 6.4 \times 10^6$ m at equator

At equator $\frac{F_c}{F_{GR}} = \frac{m\omega^2 r}{mg} = \frac{\omega^2 r}{g} = \frac{(2\pi)^2 (6.4 \times 10^6)}{(86400)^2 \times 9.81} \approx 3.5 \times 10^{-3}$

NP + SP are on the rotational axis of Earth

$r = 0$, $F_c = 0$, $\frac{F_c}{F_{GR}} = 0$

F_c/F_{GR} is independent of m of body

Activity 8 **(30 minutes)**

Roche model

You should now read Section 4.3 of FKR. Unless you are quite familiar with vector derivatives you will certainly trip over the Euler equation (FKR Equation 4.4) and the Roche potential (FKR Equation 4.5). Just take them on trust for now, we will come back to them in our discussions and questions below.

Keywords: **Roche potential**, **detached binary**, **contact binary**, **ellipsoidal variation**, **Euler equation**, **inner Langrangian point** ■

KEPLERIAN ORBITS

Kepler's laws describe the orbital motion of two point masses that are subject only to their mutual gravitational attraction. The point masses are said to execute a Kepler (or **Keplerian**) **orbit**. Sometimes it is implicitly assumed that these orbits are circular. In practice, the orbital motion of real astrophysical systems can often be approximated very well by Kepler orbits, even though the stars involved might be extended objects, and the orbit might be perturbed by the presence of other bodies.

A key element in this reading, the **Euler equation**, is simply an expression for **Newton's second law** for a continuous fluid. Specifically, the Euler equation calculates the acceleration of a small mass element in the fluid. This acceleration is either due to an external force, or due to a pressure gradient across the fluid element. The pressure gradient acceleration is given by

$$\boldsymbol{g}_{\text{pressure}} = -\frac{1}{\rho}\nabla P \qquad (7)$$

where P and ρ denote the gas pressure and density, respectively. This generalizes the form you met in Section 1.2.2 of Block 2 in the context of hydrostatic equilibrium (see, for example, Phillips Equation 1.3). Please note that the pressure P is a *scalar* quantity, hence the function $P(x, y, z)$ which describes how pressure varies with the spatial coordinates (x, y, z) is a scalar field. In contrast, the gradient $\boldsymbol{\nabla} P$ is a *vector*, and the function $\boldsymbol{\nabla} P(x, y, z)$ is a vector field. The direction of $\boldsymbol{\nabla} P$ is perpendicular to surfaces of constant pressure.

Example 4

Derive Equation 7. (Consult Figure 26 overleaf.)

Solution

Consider the acceleration that the matter in the small cube (a 'volume element') with volume V experiences as a result of a gradient of the pressure across the volume (Figure 26). We choose the orientation of the cube such that the two surfaces A_1 and A_2 coincide with surfaces of constant pressure P_1 and P_2, and we choose a coordinate system such that the x-axis points from A_1 to A_2. Let us furthermore assume that $P_1 < P_2$.

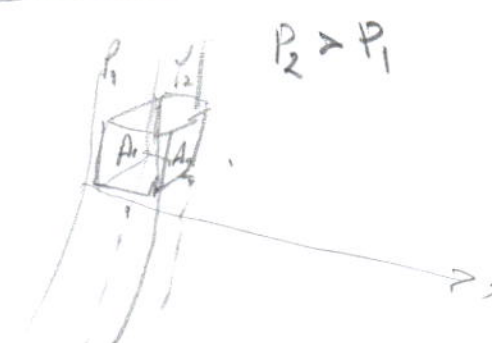

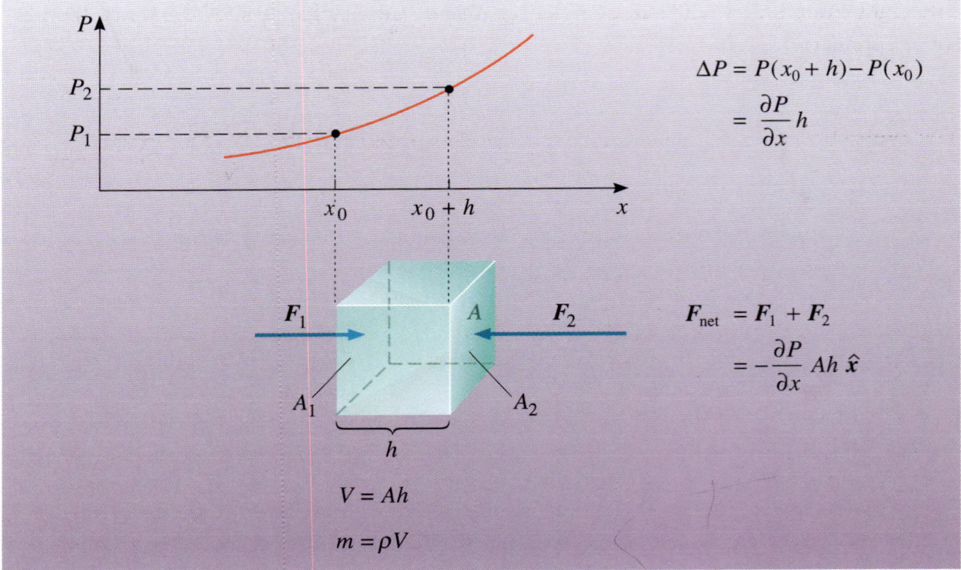

Figure 26 The acceleration due to a pressure gradient.

To calculate the acceleration of the volume element we need to divide the net force $\boldsymbol{F}$ on the element by the mass m of the element,

$$\boldsymbol{g}_{\text{pressure}} = \boldsymbol{F}/m \tag{8}$$

We calculate the mass first, then the net force.

The mass of the matter in the volume element is just m = density × volume, i.e.

$$m = \rho A h \tag{9}$$

Here A is the surface area of the surface A_1 (and A_2).

The net force on the volume element is the difference between the pressure forces on the sides A_1 and A_2. The pressure force on A_1 is in the positive x-direction, the pressure force on A_2 is in the negative x-direction. As 'magnitude of force = pressure × surface area' we have for the magnitude of the net force

$$|\boldsymbol{F}| = P_2 \times A - P_1 \times A = (P_2 - P_1) \times A$$

As $P_1 < P_2$ the net force must be in the negative x-direction, i.e. we have

$$\boldsymbol{F} = -(P_2 - P_1)A\hat{\boldsymbol{x}} \tag{10}$$

where $\hat{\boldsymbol{x}}$ is a vector of unit length in positive x-direction.

Because of our choice of the geometry and frame of reference the gradient $\boldsymbol{\nabla}P$ of the pressure points in the positive x-direction. It can be written as

$$\boldsymbol{\nabla}P = \frac{\mathrm{d}P}{\mathrm{d}x}\hat{\boldsymbol{x}} = \lim_{h\to 0}\frac{P_2 - P_1}{h}\hat{\boldsymbol{x}}$$

If the volume element is small enough we can write

$$\boldsymbol{\nabla}P \approx \frac{P_2 - P_1}{h}\hat{\boldsymbol{x}}$$

Multiplying both sides by $(-A \times h)$ gives

$$-Ah\boldsymbol{\nabla}P \approx -(P_2 - P_1)A\hat{\boldsymbol{x}}$$

The right-hand side is the same as the right-hand side of Equation 10. Hence the net force can also be written as

$$\boldsymbol{F} = -Ah\boldsymbol{\nabla}P \tag{11}$$

The only step left is to insert Equations 11 and 9 into Equation 8. This gives

$$\boldsymbol{g}_{\text{pressure}} = \frac{-Ah\boldsymbol{\nabla}P}{\rho Ah}$$

which simplifies to Equation 7.

Note that the net force in y- and z-direction is zero, as the corresponding pressure forces on opposite sides of the cube cancel to zero. ■

FKR Equation 4.4 reads

$$\frac{\partial \boldsymbol{v}}{\partial t} + (\boldsymbol{v}\cdot\boldsymbol{\nabla})\boldsymbol{v} = -\boldsymbol{\nabla}\Phi_{\text{R}} - 2\boldsymbol{\omega}\times\boldsymbol{v} - \frac{1}{\rho}\boldsymbol{\nabla}P \tag{12}$$

and is a variant of the Euler equation for stellar plasma in the co-rotating binary frame. The right-hand side has three terms. The first, written as a (vector) gradient $\boldsymbol{\nabla}\Phi_{\text{R}}$ of a scalar function, the Roche potential Φ_{R}, combines the acceleration due to the gravitational force from star 1, the gravitational force from star 2 and the centrifugal force. The second term is the acceleration due to the Coriolis force. (Note that the symbol '$\wedge$' used by FKR denotes the vector cross product $\times$.) The third term is the acceleration due to the gas pressure gradient. Do not worry about the left-hand side – this complicated form of the acceleration $\mathrm{d}\boldsymbol{v}/\mathrm{d}t$ arises because the velocity of the fluid element changes as it moves along with the fluid. We will come back to this later in the Study Guide.

Question 13

In the expression for the Roche potential Φ_{R}, FKR Equation 4.5, identify and explain the functional form of the three terms on the right-hand side. ■

So why are we so obsessed with the Roche potential? Because the Euler equation tells us that in equilibrium, for negligible fluid flow velocities, the surfaces of constant Roche potential, the Roche equipotentials, are also surfaces of constant pressure.

- ■ Why is this so?
- ☐ If the velocities vanish ($v = 0$) the Euler equation reduces to $\boldsymbol{\nabla}\Phi_{\text{R}} = (1/\rho)\boldsymbol{\nabla}P$. In words: the vector gradients of Roche potential and pressure are parallel. Quite generally, the gradient of a scalar quantity is everywhere perpendicular to the surface on which this quantity is constant. Hence the surface of constant pressure and the surface of constant Roche potential coincide! ■

In particular, the *surface* of a star in a binary coincides with a Roche equipotential, hence the shape of the Roche equipotential determines the *shape* of the star.

Question 14

The *Roche model* for stars in a binary makes three main assumptions. State them. ■

Use the Roche animation to familiarize yourself with the shape of Roche equipotentials.

Activity 9 **(30 minutes)**

Roche animation

From *The Energetic Universe* MM guide, start up the animation sequence 'Roche animation'. There are two modes of operation. In the first you view the potential in slices parallel to the orbital plane. In the second mode, you pick a value for the Roche potential, and the corresponding equipotential surface is animated and viewed from different angles, to allow you to appreciate its full three-dimensional structure.

Keywords: none ■

Question 15

The L_1 point is a saddle point of the Roche potential. Describe the characteristics of a saddle point (see also Figure 27). ■

Figure 27 A familiar surface with a saddle point.

Question 16

How many extrema (points where $\nabla \Phi_R = 0$) does the Roche potential have? Use the animation sequence (Activity 9) to find out. ■

It is obvious from, for example, Activity 9, that the Roche-lobe filling star has a non-spherical shape. If this star is a major contributor to the binary star's total emission the light curve shows an additional modulation on the orbital period. These so-called ellipsoidal variations arise as the star is seen from different viewing angles throughout the orbital period. By comparing the light curve shape with a model of how the emerging flux is distributed over the surface of the donor star it is sometimes possible to determine the binary inclination.

2.2 Hanging out with friends: binary star evolution

Mass transfer is the prerequisite of mass accretion and the plethora of high-energy phenomena that accompany accretion. Mass transfer does not just happen; there is a cause, a physical mechanism driving it. As we have already mentioned above, there are two chief ways in which a long lasting phase of mass transfer by Roche-lobe overflow could come about. In the language of the schematic diagram shown in Figure 21 the mass donor either has to keep expanding, or its Roche lobe has to be squeezed continuously in order to maintain a steady mass overflow.

Stars do indeed expand during their life. Figure 28 shows the radius evolution of a single star with mass $5M_\odot$. There are three different epochs characterized by a significant radius increase. The first epoch of gentle growth (between the points labelled A and B in the figure) is the core hydrogen-burning phase. The second epoch, with a much more rapid growth (C–E), is the phase of expansion towards the giant branch and along the first giant branch, but before ignition of core helium burning (at E). The third phase (F–I) is the expansion along the asymptotic giant branch after termination of core helium burning. Mass transfer driven by the expansion of the donor star in one of these three phases is usually called case A, case B or case C mass transfer, respectively.

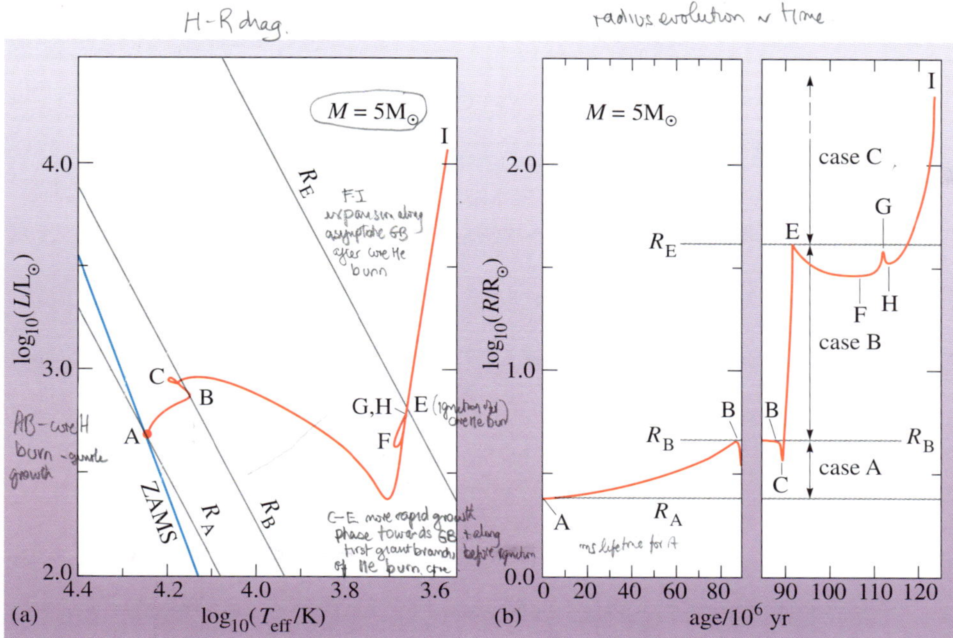

Figure 28 (a) The evolutionary track of a star with mass $5M_\odot$ in the H–R diagram, and (b) the corresponding radius evolution of this star as a function of time.

The difference between these cases is immediately obvious from Figure 28. First, the radii of potential donor stars in a case C mass transfer are larger than in a case B, and these in turn are larger than in a case A. A larger radius implies that the orbital separation must be larger to accommodate the donor star. Then Kepler's law tells us that the orbital period must be longer as well. So case A mass transfer binaries have short orbital periods, case B mass transfer binaries have intermediate orbital periods, and case C binaries have long orbital periods when mass transfer starts.

A second difference is due to the different rates of radius increase in the different phases. For case A, the radius grows on the nuclear time $t_{\rm nuc}$ of the donor star (the main-sequence lifetime), while in case B the radius increase is on the donor star's thermal time $t_{\rm th}$. The thermal time is effectively the so-called Kelvin–Helmholtz time $t_{\rm KH} = GM^2/RL$ (where M, R and L are the mass donor's mass, radius and luminosity, respectively). As $t_{\rm th}$ is typically a factor 1000 shorter than $t_{\rm nuc}$ (Block 2, Section 2.4) the expected rate at which mass is transferred must be larger by about this factor for case B than for case A. For case C evolution the radius increase and hence the transfer rate is set not by the thermal time, but by the growth of the core mass of the star.

Note that the Kelvin–Helmholtz timescale is given the symbol $\tau_{\rm KH}$ in Block 2. We have used $t_{\rm KH}$ here to align with FKR.

THE MASS TRANSFER RATE

The secondary star is the Roche-lobe filling component of the semidetached binary. The mass of the secondary is M_2. As a result of Roche-lobe overflow this mass decreases, so the derivative dM_2/dt is negative. The **mass transfer rate**, the rate at which matter flows from the secondary into the primary's Roche lobe, is usually taken to be positive. Therefore the mass transfer rate is just $(-dM_2/dt) > 0$. Following the dot notation we have

$$-\frac{dM_2}{dt} \equiv -\dot{M}_2$$

You will calculate typical mass transfer rates in the following example.

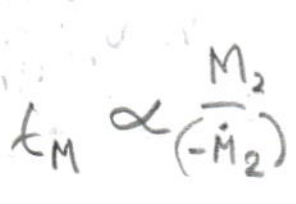

Example 5

Estimate the mass transfer rate $-\dot{M}_2$ for a donor star in a case A and case B mass transfer phase, assuming that the mass transfer timescale $t_M = M_2/(-\dot{M}_2)$ is roughly given by the nuclear and thermal time of the donor star, respectively. Consider a $1M_\odot$ and $5M_\odot$ star.

Solution

The nuclear timescale is just the main-sequence lifetime of the stars. This is roughly

$$t_{nuc} = t_{MS} \approx 10^{10}\,\text{yr}\left(\frac{M_2}{M_\odot}\right)^{-3}$$

(assuming that the main-sequence lifetime of the Sun is 10×10^9 yr, and that the luminosity of main-sequence stars scales with mass as M^4; but see also Question 23 of Block 2). For a case A system, we have $t_M \sim t_{MS}$. So as

$$t_M \equiv \frac{M_2}{-\dot{M}_2}$$

we also have

$$t_{MS} \approx \frac{M_2}{-\dot{M}_2}$$

hence $$\frac{M_2}{-\dot{M}_2} \approx 10^{10}\,\text{yr}\left(\frac{M_2}{M_\odot}\right)^{-3}$$

Solving this for the transfer rate gives

$$-\dot{M}_2(\text{case A}) \approx \frac{M_2}{t_{MS}} \approx M_2\left(10^{10}\,\text{yr}\left(\frac{M_2}{M_\odot}\right)^{-3}\right)^{-1}$$

or

$$-\dot{M}_2(\text{case A}) \approx \left(\frac{M_2}{M_\odot}\right)^4 \times 10^{-10}M_\odot\,\text{yr}^{-1} \tag{13}$$

So the approximate case A transfer rate for $M_2 = 1M_\odot$ is $10^{-10}M_\odot\,\text{yr}^{-1}$, while for $M_2 = 5M_\odot$ the transfer rate is about $5^4 \times 10^{-10}M_\odot\,\text{yr}^{-1} \approx 6 \times 10^{-8}M_\odot\,\text{yr}^{-1}$.

The thermal time t_{th} is just the Kelvin–Helmholtz time. This typically is 1/1000 of the nuclear time, so the corresponding transfer rate $-\dot{M}_2 = M_2/t_{th}$ must be about 1000 times larger than the case A rate. Hence the case B transfer rate is approximately

$$-\dot{M}_2(\text{case B}) \approx \left(\frac{M_2}{M_\odot}\right)^4 \times 10^{-7}M_\odot\,\text{yr}^{-1} \tag{14}$$

So we expect $10^{-7}M_\odot\,\text{yr}^{-1}$ if $M_2 = 1M_\odot$, and $6 \times 10^{-5}M_\odot\,\text{yr}^{-1}$ if the mass is $5M_\odot$. ■

Question 17

Express these transfer rates in cgs units. ■

In phases where the radius expansion due to nuclear evolution is very slow, (e.g. for main-sequence stars with mass $\lesssim 1M_\odot$), any mass transfer would proceed at a very, very slow rate. In these cases another driving mechanism may dominate, the loss of orbital angular momentum. Here it is not the donor star expansion which maintains mass overflow, but a continuous shrinking of the donor's Roche lobe, caused by fascinating mechanisms such as the emission of **gravitational waves** and **magnetic stellar wind braking**. You will learn more about these, and about how to calculate the mass transfer rate more accurately for a given binary star, in Activity 10.

Activity 10 (1 hour)

Binary evolution

You should now read Section 4.4 of FKR. As before, focus on the physical concepts conveyed by the text. Do not spend too much time on trying to derive equations given in the text. The detailed discussion below complements this reading, and highlights the important equations.

Keywords: **orbital angular momentum**, **conservative mass transfer**, **unstable mass transfer**, **Roche lobe radius** ■

Note that FKR uses the 'ln' notation for natural logarithms and the 'log' notation for log to the base 10.

■ Define the term *Roche lobe radius*.

□ The Roche lobe radius is the radius of a sphere that has the same volume as the Roche lobe. ■

Activity 11 (30 minutes)

Approximations for the Roche lobe

Use the spreadsheet to check how well two approximate expressions for the Roche lobe radius, FKR Equations 4.6 and 4.7, agree with each other. Consider mass ratios between 0.1 and 1.5, and calculate the relative difference $\Delta(R_2/a)/(R_2/a)$.

Keywords: none ■

A rather remarkable property of Roche-lobe filling stars is that their mean density is effectively fixed by the orbital period. We have

$$P_{\rm hr} \cong \frac{10.5}{\sqrt{\bar{\rho}/{\rm g\,cm^{-3}}}} \quad \text{(rearranged FKR 4.10)} \quad (15)$$

Note that the symbol $P_{\rm hr}$ reads 'orbital period, measured in hours', i.e. $P_{\rm hr} = (P_{\rm orb}/\text{hour})$.

Question 18

(a) Use Kepler's law and Paczyński's approximation for the Roche lobe radius (FKR Equation 4.7) to derive Equation 15. (*Hint*: Remember $\bar{\rho} = M_2/(4/3)\pi R_2^3$)

(b) Which characteristic timescale is $\propto (G\rho)^{-1/2}$?

(c) For what range of mass ratios is Equation 15 valid? ■

The radius of low-mass main-sequence stars scales roughly as the mass, $R_* \propto M_2$. The mean density therefore scales as $M_2/R_*^3 \propto M_2^{-2}$. Hence in this case the period–mean density relation (Equation 15) simplifies to the very useful rule of thumb

Handwritten top margin: R_x in low mass ms. scale roughly with M ∴ $R_* \propto M_2$, mean density is therefore scales as $\bar{\rho} \propto \dfrac{M_2}{R_2^3} = \dfrac{M_2}{M_2^3} = M_2^{-2}$

Left margin handwriting:
$\bar{\rho} \propto M_2^{-2}$
$(10.5)^2 \propto M_2^{-2}$ eq.15
$\dfrac{P_{hr}^2}{M_2} \propto (10.5)^2$
$M_2 \propto \dfrac{P_{hr}}{10.5} \sim 0.095\,P_{hr}$
$\dfrac{M_2}{M_0} = m_2 = 0.11\,P_{hr}$

Cataclysmic variable — compact binary systems consisting of a WD + low M ms. in which optical emission comes from accretion disc around WD (primary)

$$\frac{M_2}{M_0} = m_2 = 0.11 P_{\text{hr}} \qquad (m_2 \text{ the mass of star 2}) \qquad \text{(FKR 4.11) (16)}$$

Remember, FKR uses the notation $m_2 = M_2/M_\odot$, i.e. m_2 is the secondary mass measured in solar mass units.

Example 6

(a) What is the approximate orbital period of a cataclysmic variable (CV) with a M3 main-sequence donor star? (b) What is the orbital period if this M3 star is the donor star in a low-mass X-ray binary with a black hole of mass $10M_\odot$? (*Hint*: The effective temperature of a M3 main-sequence star is roughly 3300 K.)

Solution

(a) To answer this question we need to know the mass of the donor star. From Figure 49 in Block 1 we see that the mass of a main-sequence star with effective temperature 3300 K is approximately $0.5M_\odot$. Using Equation 16 we find

$$P_{\text{hr}} = \frac{m_2}{0.11} = \frac{0.5}{0.11} = 4.9$$

i.e. the orbital period is 4.9 h. The relation between mass and orbital period we used here is based on the orbital period–mean density relation (Equation 15). This relation is a good approximation as long as the mass ratio $q = M_2/M_1$ is smaller than unity. In a CV the primary is a white dwarf, and the typical mass of a white dwarf is $0.6M_\odot$. So the mass ratio is indeed less than unity. *[margin: $q = \frac{M_2}{M_1} < 1$]*

(b) If the primary is a black hole with mass $10M_\odot$, the mass ratio is 0.05 and all relations we used are still valid. So the low-mass X-ray binary has the same orbital period, 4.9 h. *[margin: $q = 0.05$]*

A word of caution applies. In many low-mass X-ray binaries the donor star appears to be *undermassive*, i.e. has a mass that is much smaller than the rule of thumb – Equation 16 – would suggest. ■ *[$m_2 = 0.11 P_{hr}$]*

Left margin handwriting: (NOTE) donor R_* and Roche lobe radius must move in step — the key to understanding how mass transfer rate adjusts to changes in orbital separation

The key to understanding how the mass transfer rate adjusts to changes in the orbital separation and the donor's size is the insight that the donor's stellar radius R_* and Roche lobe radius R_2 must *move in step*. (Note that in Section 1.2 we have used the symbol R_* to denote the radius of the accretor, while here R_* refers to the radius of the donor star. In general, the meaning will always be clear from the context – so be alert!) In the language of Figure 21 this is rather obvious. The star is always just as big as its Roche lobe – if it were bigger, the surplus matter would quickly flow off to the primary star. A more formal way to express this is by FKR Equation 4.19,

Left margin handwriting: the transfer rate at any instant is sensitive to the difference between stellar radius + Roche lobe radius

$$-\dot{M}_2(\text{inst}) = \dot{M}_0 \exp\left(\frac{R_* - R_2}{H_*}\right)$$

[handwritten labels: stellar radius difference / photosphere scale height / $\dot{M}_2(\text{inst})$]

($\dot{M}_0 \approx 10^{-8} M_\odot\,\text{yr}^{-1}$), which shows that the instantaneous transfer rate is a sensitive function of the difference between Roche lobe radius and stellar radius (Figure 29).

The quantity H_* that appears in FKR Equation 4.19 is the photospheric **scale height** and is a measure of how *sharp* the stellar rim is. It is defined as the length scale over which the stellar pressure P in the outermost layers of the star, close to the stellar photosphere (where $P = P_{\text{Ph}}$), drops by a factor $e \approx 2.7$ (Figure 30).

LOGARITHMIC DERIVATIVES

A rather convenient way to work out the derivative of a product or quotient is the use of logarithmic derivatives (see also Block 1, Section 3.6). For obvious reasons the derivative of the form

$$\frac{\dot{a}}{a} = \frac{\mathrm{d}}{\mathrm{d}t}(\log_e a)$$

is called the **logarithmic derivative** of a. If a quantity a is given as $a = b^n/c^m$, where n and m are positive numbers, then the logarithmic derivative of a is

$$\frac{\dot{a}}{a} = n\frac{\dot{b}}{b} - m\frac{\dot{c}}{c}$$

This is a simple consequence of the chain and product rule:

$$\dot{a} = \frac{\mathrm{d}}{\mathrm{d}t}(b^n c^{-m}) = [nb^{n-1}\dot{b} \times c^{-m}] + [b^n \times (-m)c^{-m-1}\dot{c}]$$

$$= n\frac{b^n}{bc^m}\dot{b} - m\frac{b^n}{cc^m}\dot{c} = n\frac{a}{b}\dot{b} - m\frac{a}{c}\dot{c}$$

which reproduces what we have written above when we divide both sides by a. This can be generalized to expressions with more factors in the numerator and denominator. For each factor in the numerator add the logarithmic derivative of this factor, for each factor in the denominator subtract the logarithmic derivative of this factor.

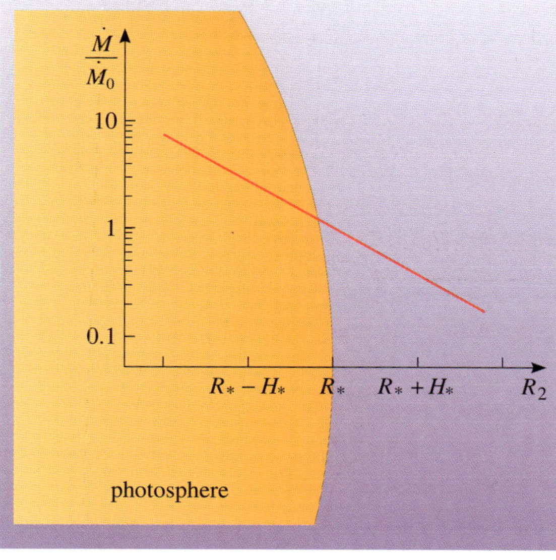

Figure 29 Mass transfer rate, in units of the rate $\dot{M}_0$, as a function of Roche lobe radius R_2. Note that the $\dot{M}$-axis is in logarithmic units. The large circle indicates the position of the secondary star's photosphere ('stellar surface').

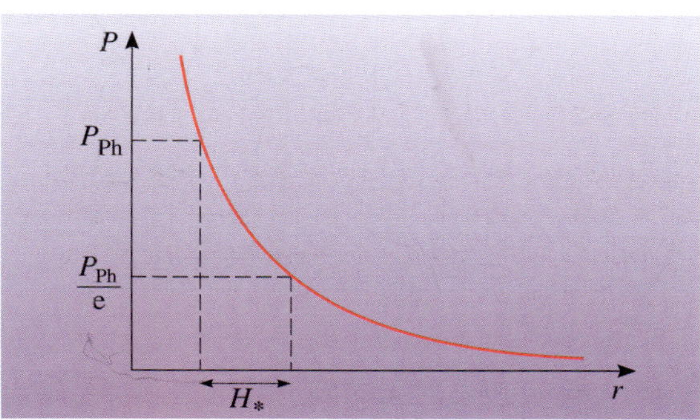

Figure 30 Photospheric pressure scale height.

Handwritten (top left):

$-\dot{M}_2(\text{inst}) = \dot{M}_0 \exp\left(\frac{R_* - R_2}{H_*}\right)$

If $R_* - R_2$ changes by $\sim 2 \times H_*$ the transfer rate $-\dot{M}$ changes by only an order of mag.

If $R_* - R_2$ is initially ~ 0 then

$-\dot{M}_2(\text{inst}) \sim \dot{M}_0 \exp(0)$

If $R_* - R_2$ is $2.3 \times H_*$ then

$-\dot{M}_2(\text{inst}) = \dot{M}_0 \exp 2.3$
$= 10 \dot{M}_0$

(i.e. an order of mag. change)

For low mass m.s. stars their photosphere scale height is tiny $0.01\% R$
So. $R_* \sim R_2$
so the radii move in step

Any estimate of mass transfer rate in a semi detached binary system depends on $\frac{\dot{R}_*}{R_*} - \frac{\dot{R}_2}{R_2}$

\dot{R}_*, \dot{R}_2 can be calculated either
① identifying the time scale for the change ($t_{nuc}, t_{th} \cdots$)
② making sure donor follows a certain M, R relationship

Rate of change of Roche lobe R is found by considering the orbital angular momentum. (FKR discusses conservative mass transfer)

Conservative mass transfer = form of mass transfer in a binary system where the total mass & total angular momentum of the system remains constant

Assume
Each small mass element ΔM removes therefore ΔJ of orbital angular momentum so the wind carries same angular momentum per unit mass as the primary star due to orbital motion

ζ is the Greek letter zeta.

eq 17 $\frac{\dot{R}_*}{R_*} = \frac{\dot{R}_2}{R_2}$

Main body:

Handwritten top: $(R_ - R_2)$... $2 \times H_*$*

If the difference between R_* and R_2 changes only by about twice the value of H_* the transfer rate will change by an order of magnitude. *ie $(e^2 \sim 8) \sim 10$*

- Why is this the case?

- Suppose that initially the difference $R_* - R_2$ is 0. Then the transfer rate is $-\dot{M}_2(\text{inst}) = \dot{M}_0 \exp(0) = \dot{M}_0$. *$e^0 = 1$* If the new difference is $2.3 \times H_*$ the new transfer rate is

$$-\dot{M}_2(\text{inst}) = \dot{M}_0 \exp(2.3) \approx 10 \times \dot{M}_0 \quad \blacksquare$$

order of mag change *($e^{2.3} \sim 10$)*

Low-mass main-sequence stars have *very sharp* edges, their photospheric scale height is only a tiny 0.01% of the stellar radius. *(h^*)* So in practice we always have $R_* \approx R_2$. As the radii R_* and R_2 move in step we can write

$$\frac{\dot{R}_*}{R_*} = \frac{\dot{R}_2}{R_2} \tag{17}$$

Handwritten in box: This expression is the foundation for any estimate of the mass transfer rate in a semi detached binary system. \dot{R}_ can be calculated by either Identifying a characteristic time scale (t_{nuc}, t_{th}) or the donor has a $M \sim R$ relationship. \dot{R}_2 — consider orbital angular momentum.*

Note

This expression is the foundation for any estimate of the mass transfer rate in a semidetached binary system. The rate of change of the stellar radius can be calculated by either identifying the characteristic timescale for this change (e.g. the nuclear time, the thermal time), or by requiring that the donor follows a certain mass–radius relation. The rate of change of the Roche lobe radius can be found by considering the orbital angular momentum. FKR discusses the case of **conservative mass transfer**, where both the orbital angular momentum and the total binary mass are constant throughout the evolution. The following rather long example on isotropic wind losses retraces these steps in a slightly different context.

ans to example in different context

Example 7

Following the steps in FKR for conservative mass transfer, derive an expression for the mass transfer rate in a system where the primary star loses mass into space in the form of an isotropic stellar wind, at the same rate as it accretes mass from the secondary (see the schematic diagram in Figure 31). Assume that each small mass element ΔM in the wind removes an amount $\Delta J = (M_2/M_1)(J/M)\Delta M$ of the orbital angular momentum J. (One can show that in this case the wind carries the same angular momentum per unit mass as the primary star due to its orbital motion.) Assume further that the secondary follows a mass–radius relation with logarithmic slope ζ (i.e. $R_* \propto M_2^{\zeta}$, so that $\log_{10} R_* = \zeta \log_{10} M_2 + \text{constant}$).

$R_ = \text{constant} \, M_2^{\zeta}$ taking log of both sides*

Solution

We need to express \dot{R}_*/R_* and \dot{R}_2/R_2 in terms of the mass transfer rate, insert these in Equation 17, and solve for the transfer rate. From $R_* \propto M_2^{\zeta}$ we have simply

$$\frac{\dot{R}_*}{R_*} = \zeta \frac{\dot{M}_2}{M_2} \tag{18}$$

(see the box on 'Logarithmic derivatives'). To calculate \dot{R}_2/R_2 we note that the primary mass M_1 is constant as the primary loses as much matter in a wind as it gains through accretion, so $\dot{M}_1 = 0$. The total binary mass $M = M_1 + M_2$ does change, though:

$$\dot{M} = \dot{M}_1 + \dot{M}_2 = \dot{M}_2 < 0$$

$\dot{M}_1 = 0$

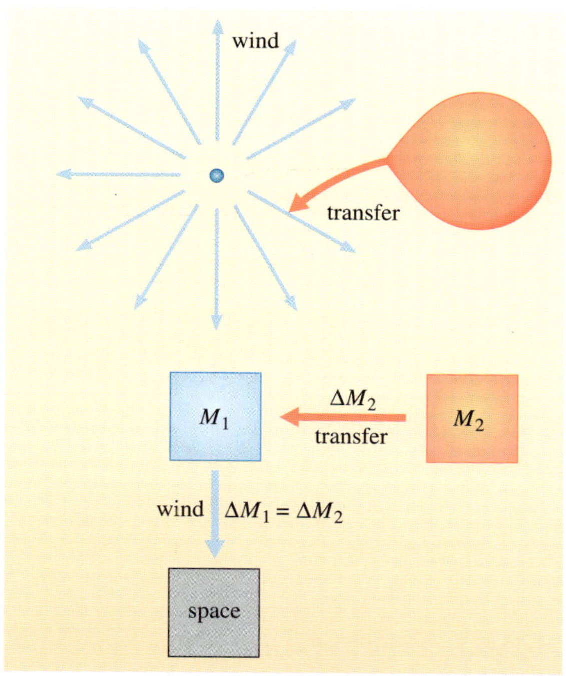

Figure 31 Schematic mass flow diagram in the presence of isotropic wind losses from the primary.

We estimate the Roche lobe radius via FKR Equation 4.7 (remember, this is fine for mass ratios smaller than unity),

$$R_2 \cong 0.462\left(\frac{M_2}{M}\right)^{1/3} a$$

The logarithmic derivative of this is

$$\frac{\dot{R}_2}{R_2} = \frac{1}{3}\frac{\dot{M}_2}{M_2} - \frac{1}{3}\frac{\dot{M}}{M} + \frac{\dot{a}}{a} \qquad (19)$$

To find the logarithmic derivative of a we use the expression for the orbital angular momentum (FKR Equation 4.13),

$$J = M_1 M_2 \left(\frac{Ga}{M}\right)^{1/2}$$

Solving for a we have

$$a = \frac{J^2 M}{G M_1^2 M_2^2}$$

hence $$\frac{\dot{a}}{a} = 2\frac{\dot{J}}{J} + \frac{\dot{M}}{M} - 2\frac{\dot{M}_1}{M_1} - 2\frac{\dot{M}_2}{M_2} = 2\frac{\dot{J}}{J} + \frac{\dot{M}}{M} - 2\frac{\dot{M}_2}{M_2} \qquad (20)$$

(remember, $\dot{M}_1 = 0$). Inserting Equation 20 into the above expression for $\dot{R}_2/R_2$ (Equation 19) gives

$$\frac{\dot{R}_2}{R_2} = \left(\frac{1}{3} - 2\right)\frac{\dot{M}_2}{M_2} + \left(-\frac{1}{3} + 1\right)\frac{\dot{M}}{M} + 2\frac{\dot{J}}{J}$$

Collecting terms gives

$$\frac{\dot{R}_2}{R_2} = -\frac{5}{3}\frac{\dot{M}_2}{M_2} + \frac{2}{3}\frac{\dot{M}}{M} + 2\frac{\dot{J}}{J} \qquad (21)$$

There are two qualitatively different types of orbital angular momentum losses. Mechanisms of the first type include gravitational radiation and magnetic braking (see the corresponding boxes). They are called *systemic* as they do not involve a (measurable) change in the binary component masses. The second type of orbital angular momentum losses arises when the binary loses mass – as in this example through a wind – and this mass carries away angular momentum.

From the information above, the amount of angular momentum ΔJ that is lost in a time Δt due to the wind can be written in terms of the mass ΔM that is lost in the time Δt:

$$\frac{\Delta J}{\Delta t} = \frac{M_2}{M_1}\frac{J}{M}\frac{\Delta M}{\Delta t}$$

In the limit $\Delta t \rightarrow 0$ we can replace $\Delta M/\Delta t$ by $\mathrm{d}M/\mathrm{d}t = \dot{M}$ and $\Delta J/\Delta t$ by $\dot{J}_{\text{wind}}$. Hence

$$\dot{J}_{\text{wind}} = \frac{M_2}{M_1}\frac{J}{M}\dot{M}$$

or

$$\frac{\dot{J}_{\text{wind}}}{J} = \frac{M_2}{M_1}\frac{\dot{M}}{M} \qquad (22)$$

The total orbital angular momentum loss is the sum of the systemic losses and the losses carried away in the wind. We therefore write

$$\frac{\dot{J}}{J} = \frac{\dot{J}_{\text{sys}}}{J} + \frac{M_2}{M_1}\frac{\dot{M}}{M} \qquad (23)$$

By substituting Equation 23 into Equation 21 we obtain

$$\frac{\dot{R}_2}{R_2} = -\frac{5}{3}\frac{\dot{M}_2}{M_2} + \left(\frac{2}{3} + 2\frac{M_2}{M_1}\right)\frac{\dot{M}}{M} + 2\frac{\dot{J}_{\text{sys}}}{J}$$

Now we use the fact that $\dot{M} = \dot{M}_2$ and factor $\dot{M}_2$ out, and M in. This gives

$$\frac{\dot{R}_2}{R_2} = \left(-\frac{5}{3} + \frac{2M_2}{3M} + 2\frac{M_2^2}{M_1 M}\right)\frac{\dot{M}_2}{M_2} + 2\frac{\dot{J}_{\text{sys}}}{J}$$

Equating this to $\dot{R}_*/R_*$ (Equation 18) we have

$$\zeta\frac{\dot{M}_2}{M_2} = \left(-\frac{5}{3} + \frac{2M_2}{3M} + 2\frac{M_2^2}{M_1 M}\right)\frac{\dot{M}_2}{M_2} + 2\frac{\dot{J}_{\text{sys}}}{J}$$

or

$$\frac{\dot{M}_2}{M_2}\left(\zeta + \frac{5}{3} - \frac{2M_2}{3M} - 2\frac{M_2^2}{M_1 M}\right) = 2\frac{\dot{J}_{\text{sys}}}{J}$$

hence finally

$$\frac{-\dot{M}_2}{M_2} = \frac{-\dot{J}_{\text{sys}}/J}{\dfrac{\zeta}{2} + \dfrac{5}{6} - \dfrac{M_2}{3M} - \dfrac{M_2^2}{M_1 M}} \qquad (24)$$

This equation allows one to calculate the mass transfer rate in a semidetached binary if the systemic angular momentum loss rate is known. The only difference between this expression for the mass transfer rate and the corresponding one for conservative mass transfer (FKR Equation 4.17) is the functional form of the denominator. You will investigate the consequences of this difference in Activity 12. ■

Congratulations if you made it this far through all this algebra!

GRAVITATIONAL WAVE RADIATION

One of the most dramatic predictions of Einstein's theory of **general relativity** is the existence of gravitational waves. In general relativity, space and time make up the unified four-dimensional spacetime. Mass curves spacetime. A variation of the mass distribution in a system causes a corresponding variation in the spacetime curvature. These changes can propagate outwards in the form of gravitational waves, carrying away energy and angular momentum. The 'ripples' in spacetime travel with the speed of light (Figure 32).

For a binary star with component masses m_1 and m_2 (in solar mass units) in a circular orbit with orbital period P_{hr} (in hours) the loss rate $\dot{J}_{\mathrm{GR}}$ of orbital angular momentum J is given by

$$\frac{\dot{J}_{\mathrm{GR}}}{J} = -1.27 \times 10^{-8} \frac{m_1 m_2}{(m_1 + m_2)^{1/3}} P_{\mathrm{hr}}{}^{-8/3} \mathrm{yr}^{-1}$$

The emission of gravitational waves leads to a slow spiral-in of the binary. The predictions made by the more general form of this *quadrupole formula* for elliptical orbits have been beautifully confirmed to very high accuracy by observations of the orbital decay of the Hulse–Taylor binary pulsar (see Activity 13).

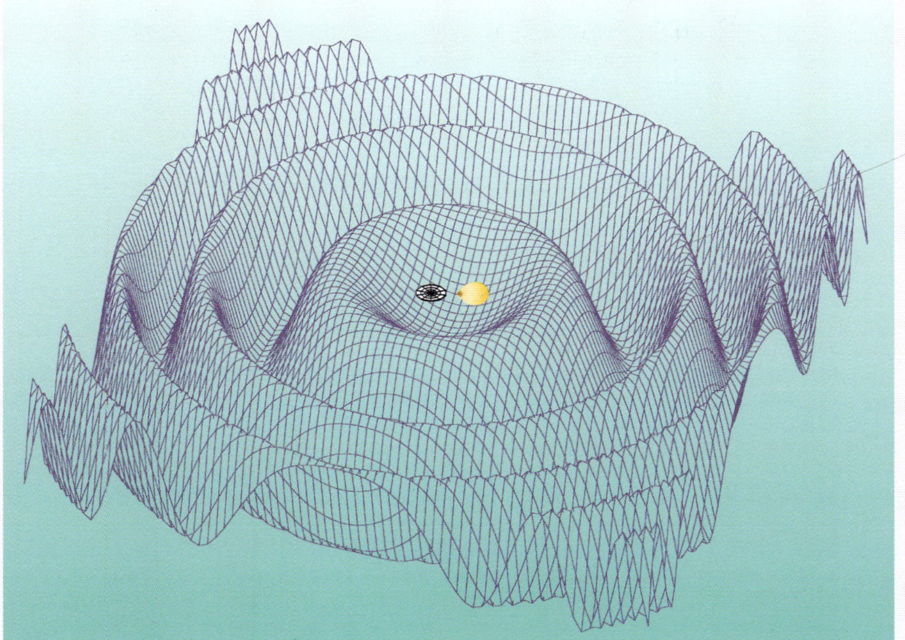

Figure 32 An illustration of gravitational waves emitted by a binary system.

MAGNETIC BRAKING

Observations of CVs with orbital periods longer than about 3 h suggest that there is an orbital angular momentum loss mechanism at work with 10 to 100 times higher efficiency than gravitational waves. Most researchers believe that this is magnetic stellar wind braking, the same effect that causes the observed gradual spin-down of single main-sequence stars. Such stars have a stellar wind, a constant flow of stellar plasma away from the star, *and* a large-scale magnetic field that essentially corotates with the star – as the magnetic field of a rotating bar magnet would. (You will learn more about magnetic fields and their interaction with plasma in Section 7.) Through interaction with the magnetic field the material in the wind acquires a large specific angular momentum (angular momentum per unit mass). Thus the braking is very efficient, although there is very little mass lost in the wind. **Tidal forces** ensure that the secondary in CVs is rotating synchronously with the orbit, i.e. the secondary is tidally locked to the orbit. Hence any loss of rotational angular momentum leads to a net loss of orbital angular momentum.

Activity 12 (45 minutes)

Mass transfer stability

Clearly, for Equation 24 to work the denominator has to be positive. (Note that both the numerator and the left-hand side are positive.) Otherwise $\dot{M}_2$ would come out positive, which is clearly nonsensical if the star is supposed to lose mass. For the standard case $\zeta = 1$, find a condition on the mass ratio for the denominator to be positive. To make it easier, break it down into the following steps:

(a) Rewrite the denominator on the right-hand side of Equation 24 in terms of q.

(b) Then use a spreadsheet to calculate the denominator for a number of q values in the range 0.1 to 2. Plot the result.

(c) For what values of q is the denominator positive?

(d) Compare your result with the corresponding condition one can derive from FKR Equation 4.17.

We discuss the implication of this 'stability denominator' for mass transfer stability below.

See the section 'Comments on activities' for a model answer.

Keywords: none

In Activity 12 you found that the denominator that appears in FKR Equation 4.17 or Equation 24 is positive as long as the mass ratio is less than some critical value of order unity. If the mass ratio is larger and therefore the denominator negative these equations give a non-physical value for the mass transfer rate $(-\dot{M}_2)$ - a negative rate. This signals that for large enough mass ratios one of the assumptions that have been used to derive these equations must break down. It is in fact the assumption that the stellar and Roche lobe radius move in step (Equation 17). For large mass ratios the Roche lobe shrinks relative to the star, i.e. it cannot accommodate the star. Formally, the difference $R_* - R_2$ between the stellar radius and Roche lobe radius becomes positive and continues to increase. Hence, according to FKR Equation 4.19, the transfer rate continues to grow. If this is the case mass transfer is said to be unstable. Depending on the nature of the donor star this can either lead to a complete disruption of the binary, or to a transient, short-lived phase with a very high transfer rate. In the latter case the system stabilizes and reverts to the normal rate given by, for example, Equation 24 once the mass ratio is smaller than the critical limit.

■ You have investigated conservative mass transfer and mass transfer with isotropic wind losses. Which type of mass transfer is more stable?

❑ Mass transfer with wind losses is more stable than conservative mass transfer. The answer (d) in Activity 12 showed that in the case of wind losses the upper limit on the mass ratio for stable mass transfer (denominator positive) is larger than in the conservative case. ■

[Handwritten margin notes: $q < 1.3$ -- Dcons is +ve ; $q < 1.75$ Dwind is +ve +larger than Dcons.]

Activity 13 (20 minutes)

The Hulse–Taylor binary pulsar

In 1993 the astrophysicists Russell Hulse and Joseph Taylor received the Nobel Prize for Physics 'for the discovery of a new type of pulsar, a discovery that has opened up new possibilities for the study of gravitation'.

Connect to the S381 home page and access the list of Web-based activities (under Course resources). Select Block 3, Activity 13 and follow the link to the Nobel Foundation and find information on the 1993 Physics Prize. Then answer Question 19.

Keywords: none ■

[Handwritten margin notes: Activity 13 Hulse-Taylor binary pulsar. Course resources Web Site. BK3 Activity 13 follow link to Nobel Foundation. Info on 1993 Physics Prize. See in Activity Instructions book P4. Paper in My docs/s381]

Question 19

When did Hulse and Taylor discover the binary pulsar, using what instrument in which waveband? ■

[Handwritten answer: (19) Binary pulsar was discovered in 1974, with the Arecibo 300m radio telescope. Obs. made in radio band.]

Question 20

Gravitational waves are thought to be the dominant driving mechanism of mass transfer in short-period CVs. Using FKR Equation 4.17 and the equation in the box on gravitational wave radiation, estimate the mass transfer rate in a CV with a 2-hour orbital period and a white dwarf with mass $1M_\odot$. (*Hint*: You will need to estimate the mass of the secondary star from Equation 16.) ■

*[Handwritten notes:
(20) FKR eq 4.17 (mass transfer rate) $\frac{-\dot{M_2}}{M_2} = \frac{-\dot{J}/J}{\frac{4}{3} - M_2/M_1}$
$\frac{\dot{J}_{GR}}{J} = -1.27 \times 10^{-8} \frac{m_1 m_2}{(m_1+m_2)^{1/3}} P_{hr}^{-8/3}$ yr^{-1} (angular momentum loss rate)
$P = 2hr$ WD $= 1M_\odot$ $m_1 = 1$
(16)... $\frac{M_2}{M_\odot} = 0.11 P_{hr}$ $m_2 = 0.22$
Subs into eq. $\frac{\dot{J}_{GR}}{J} = -1.27 \times 10^{-8} \frac{0.22}{(1.22)^{1/3}} 2^{-8/3}$ yr^{-1}
$= -4.12 \times 10^{-10}$ yr^{-1}
$\frac{-\dot{M_2}}{0.22 M_2} = \frac{4.12 \times 10^{-10}}{\frac{4}{3} - 0.22/1}$
$\therefore |\dot{M_2}| = 8.14 \times 10^{-11} M_\odot$ yr^{-1} (|M| $= 5.13 \times 10^{15}$ g s^{-1})]*

2.3 Accretion disc formation

So far we have considered when and how mass is transferred in a semidetached binary. We now study the further fate of the stellar material once it leaves the donor star via the inner Lagrangian point. We follow the mass on its path towards its new home, the primary.

Activity 14 (45 minutes)

Stream to torus

You should now read Section 4.5 of FKR. In the second paragraph of page 59 there is a reference back to Chapter 2 of FKR which you should ignore, i.e. accept the information given as a fact. At about this point it may also help you if you jump ahead to Activity 15, watch the accretion stream animation, and then return to this reading of FKR.

[Handwritten notes: Activity 14. FKR Section 4.5. do Activity 15 after the 2nd paragraph on p 59]

Keywords: **circularization radius** ■ — *[Handwritten: the radius of a circular orbit centred on the accretor in any interacting binary system where the orbital angular momentum per unit mass is the same as the material leaving the donor at the inner Lagrangian point.]*

If you have not already done so, you should now watch the accretion stream animation in Activity 15.

Activity 15 **(20 minutes)**

The accretion stream

From *The Energetic Universe* MM guide, start the animation sequence entitled 'The accretion stream'. This shows the ballistic trajectory of the mass transfer stream that emanates from the L_1 point.

The unfortunate gas stream *collides* with itself! What fate awaits the stream? To be continued … (in Activity 19). (After watching the animation, return to Activity 14.)

Keywords: none ■

The velocity of the gas stream is of order the sound speed when it is in the vicinity of the L_1 point. As you have seen in Section 4.7 of Block 1 (Equation 147), the speed of sound in a gas, denoted by the symbol c_s, depends on the ratio of the pressure to the density of the gas, or equivalently on the temperature of the gas (since $P/\rho \propto T$). In fact, for our purposes a good approximation is

$$c_s \cong \left(\frac{P}{\rho}\right)^{1/2} \tag{25}$$

or

$$c_s \cong 10\left(\frac{T}{10^4\,\mathrm{K}}\right)^{1/2}\,\mathrm{km\,s^{-1}} \tag{26}$$

(These also appear as FKR Equations 2.20 and 2.21, i.e. in a chapter of FKR which you did not read.)

■ Does the right-hand side of Equation 25 really have the dimension of a velocity?

☐ Yes. Pressure has the dimension of force per surface area, i.e. the unit $(\mathrm{kg \times m\,s^{-2}})/\mathrm{m^2} = \mathrm{kg\,m^{-1}\,s^{-2}}$, while density has the unit $\mathrm{kg\,m^{-3}}$. Hence P/ρ has $(\mathrm{kg\,m^{-1}\,s^{-2}}) \times (\mathrm{kg\,m^{-3}})^{-1} = \mathrm{kg^{1-1}\,m^{3-1}\,s^{-2}} = \mathrm{m^2\,s^{-2}}$. The square root of this is $\mathrm{m\,s^{-1}}$, the unit of velocity. ■

Question 21

(a) Estimate the sound speed in the photosphere of a $1\mathrm{M}_\odot$ and a $0.5\mathrm{M}_\odot$ main-sequence star. (*Hint*: Estimate the photospheric temperature of these stars from Figure 49 of Block 1, and use Equation 26.)

(b) Compare these sound speeds to the orbital speed of a test body with very small mass that circles the respective star just above the stellar photosphere, i.e. on an orbit with radius equal to the stellar radius. (*Hint*: Assume that the stellar radii of the two stars are $1\mathrm{R}_\odot$ and $0.5\mathrm{R}_\odot$, respectively.) ■

The answer to Question 21 tells us that as seen from the white dwarf, the orbital velocity of the L_1 point itself is highly supersonic. The matter in the stream effectively free-falls towards the primary, i.e. it follows a **ballistic trajectory**. The effect of the gas pressure on the motion of the stream is negligible. The geometric cross–section of the stream when it crosses over into the lobe of the primary star is of order H_*R_* – recall from our discussion above, just before Equation 17, that H_* is the photospheric pressure scale height. As $H_* \ll R_*$ the stream cross-section is only a rather small fraction of the stellar surface area.

Question 22

Define the term *circularization radius*. ■

Example 8

In your last reading you found statements to the effect that *a circular orbit has the lowest energy for a given orbital angular momentum.* Consider bound orbits and convince yourself that the statement is true.

We provide all necessary information on elliptical orbits below.

Solution

In general, a test mass – a rock, say – that is gravitationally bound to a central body – a star, say – orbits this body on an elliptical orbit. The geometry of an ellipse is shown in Figure 33. There are two axes of symmetry; the semimajor axis has length a, the semiminor axis has length b. The star with mass M coincides with a focal point of the ellipse. This point is on the major axis, offset from the centre by a distance ea, where e is the **eccentricity** of the ellipse. Obviously we have $0 \leq e < 1$. It follows that the smallest distance between the rock and the star, the periastron, is just $r_\mathrm{p} = a - ae = a(1 - e)$. Similarly the largest distance between the rock and star, the **apastron**, is $r_\mathrm{a} = a(1 + e)$. You need one last bit of information to verify the above statement: the orbital speed v of the rock is given by

$$v^2 = \frac{2GM}{r} - \frac{GM}{a}$$

Here r is the instantaneous separation between the rock and star (Figure 33b).

Now consider the orbital angular momentum and total orbital energy of the rock. Let m be the mass of the rock. Then the magnitude of the orbital angular momentum is $J = rmv \sin\theta$ (with θ being the angle between the direction of r and v; see Equation 132 in Block 1, and Figure 33). It does not matter where we measure J, as J has the same value throughout the orbit. It is most easily calculated at points where the

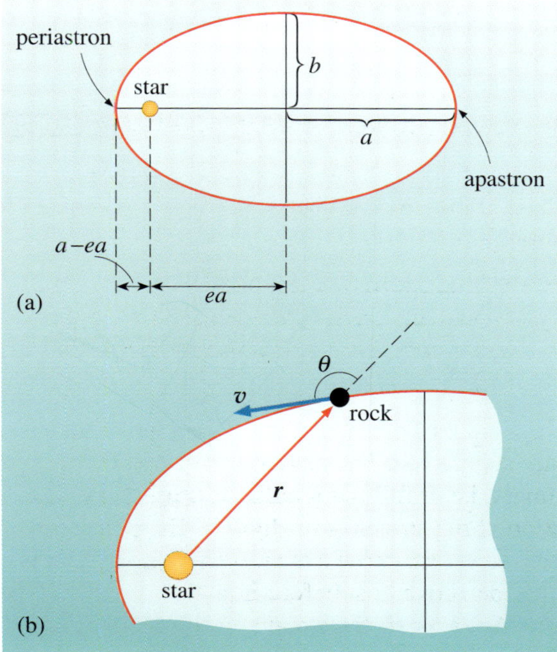

Figure 33 Elliptical orbit.

angle θ is $\pi/2$, i.e. at either the periastron or apastron, because $\sin(\pi/2) = 1$. At the periastron we have $r = r_p = a(1 - e)$ and the velocity $v = v_p$, with

$$v_p^2 = \frac{2GM}{a(1-e)} - \frac{GM}{a} = \frac{GM}{a}\left(\frac{2}{1-e} - 1\right) = \frac{GM}{a}\frac{1+e}{1-e}$$

Therefore the magnitude of the angular momentum is

$$J = mr_p v_p = ma(1-e)\sqrt{\frac{GM(1+e)}{a(1-e)}} = m\sqrt{\frac{a^2(1-e)^2 GM(1+e)}{a(1-e)}}$$

or simply

$$J = m\sqrt{GMa(1-e^2)} \qquad (27)$$

The orbital energy E is the sum of translational kinetic energy and gravitational potential energy (which is negative). Using the values for r and v at periastron again we obtain

$$E = \frac{1}{2}mv_p^2 - G\frac{Mm}{r_p} = \frac{1}{2}m\frac{GM}{a}\frac{1+e}{1-e} - \frac{GMm}{a(1-e)}$$

Factoring out $GM/[a(1 - e)]$ and then $\frac{1}{2}$ gives

$$E = \frac{GMm}{a(1-e)}\left[\frac{1}{2}(1+e) - 1\right] = \frac{GMm}{a(1-e)}\frac{1}{2}(1+e-2)$$

This finally simplifies to

$$E = \frac{GMm}{a(1-e)}\frac{1}{2}(e-1) = -\frac{GMm}{2a}$$

Obviously $E < 0$. This expresses the fact that the orbit is bound and therefore periodic. We are asked to find the orbit with the lowest (most negative) energy among all orbits with the same orbital angular momentum. Physically, this orbit will emerge as a result of dissipative processes that extract orbital energy but not angular momentum. To find the lowest energy orbit we consider changes of the orbital parameters a and e such that J remains constant. Equation 27 shows that if the product $a(1 - e^2)$ is constant, then J is also constant. How do such changes that keep $a(1 - e^2)$ constant affect E? We rewrite E conveniently as

$$E = -\frac{GMm}{2a} \times \frac{(1-e^2)}{(1-e^2)}$$

i.e.

$$E = -\frac{GMm}{2a(1-e^2)} \times (1-e^2) \qquad (28)$$

Then the first factor on the right-hand side of Equation 28 is constant during any extraction of energy that conserves J, since $J \propto a(1 - e^2)$, while E changes exclusively through the last factor $(1 - e^2)$. Obviously E is smallest when e becomes 0 (note the leading minus sign!). Now go back to Figure 33 and verify that an ellipse with eccentricity 0 is just a circle. ■

In summary: In a semidetached binary, material leaves the secondary star as a geometrically thin, highly supersonic gas stream and initially follows a ballistic trajectory. The stream eventually self-intersects, and dissipative processes through, for example, the collision of individual gas blobs convert kinetic energy of its bulk motion into heat. Some of this internal energy may be lost by radiation. The stream settles into the lowest energy configuration for a given angular momentum, and forms a ring or torus centred on the circularization radius (see Question 22).

The torus evolves further only if angular momentum is lost or redistributed. For material to move further in towards the primary, and for it to actually *accrete* onto the surface of the accretor, a torque must operate that removes angular momentum (see Block 1, Section 4.5). Before we consider the transition from a torus into an accretion disc in some detail we turn in Section 3 to the physical processes providing the necessary torques.

The name game: part 2

We suggest you skim-read Section 2, and make a list of all symbols that denote *new* variables and *new* quantities, and note down what they mean. Flag up those symbols that have (unfortunately) multiple uses.

Once you have compiled your list and explained the meaning of all entries, compare it to the one given in Appendix A5 at the end of the Study Guide.

2.4 Summary of Section 2

1. In an interacting binary system the mass donor is also referred to as the secondary, while the mass accretor is the primary. The mass ratio is $q = M_2/M_1$.
2. The maximum volume available to a binary component is the Roche lobe. The Roche lobe radius is the radius of a sphere with the same volume as the Roche lobe.
3. A binary is detached if both components reside well inside their respective Roche lobes. It is semidetached if only one component fills its Roche lobe. In a contact binary both components fill, or overfill, their Roche lobes.
4. There are two modes of mass transfer: Roche-lobe overflow in a semidetached binary, and wind accretion in a detached binary.
5. Point masses execute a Kepler orbit. Binaries are best discussed in a frame of reference that co-rotates with the orbit. The centrifugal force with magnitude

   $$F_{\rm c} = m\frac{v^2}{r} = m\omega^2 r$$

   and the Coriolis force (involving the vector product of $\boldsymbol{\omega}$ and $\boldsymbol{v}$) are pseudo-forces that occur in such a rotating frame.
6. The Euler equation is an expression for Newton's second law for a continuous fluid. One term in the Euler equation is the acceleration of a fluid element due to a pressure gradient,

   $$\boldsymbol{g}_{\rm pressure} = -\frac{1}{\rho}\nabla P$$

7. The Roche potential $\Phi_{\rm R}$ combines the acceleration due to the gravitational force from star 1, the gravitational force from star 2 and the centrifugal force. Roche equipotentials are also surfaces of constant pressure. The surface of a star in a binary coincides with a Roche equipotential.
8. The standard form of the Roche potential is valid if the two stars are centrally condensed and tidally locked to the orbit.
9. A physical driving mechanism is required to sustain stable, continuous mass transfer.

10 One driving mechanism is the expansion of the donor star due to its nuclear evolution. Depending on the evolutionary phase of the donor star we distinguish case A, case B and case C mass transfer.

For case A mass transfer a typical transfer rate is

$$-\dot{M}_2(\text{case A}) \approx \left(\frac{M_2}{\mathrm{M}_\odot}\right)^4 \times 10^{-10}\mathrm{M}_\odot\ \mathrm{yr}^{-1}$$

For case B mass transfer a typical transfer rate is

$$-\dot{M}_2(\text{case B}) \approx \left(\frac{M_2}{\mathrm{M}_\odot}\right)^4 \times 10^{-7}\mathrm{M}_\odot\ \mathrm{yr}^{-1}$$

11 Another driving mechanism for mass transfer is the loss of orbital angular momentum. This could be due to the emission of gravitational waves or due to magnetic stellar wind braking.

12 In semidetached binaries the orbital period is related to the mean density of the donor star,

$$P_{\mathrm{hr}} \cong \frac{10.5}{\sqrt{\bar{\rho}/\mathrm{g\,cm^{-3}}}}$$

13 In cataclysmic variables we have the rule of thumb $m_2 = 0.11P_{\mathrm{hr}}$.

14 The instantaneous transfer rate is a sensitive function of the difference between Roche lobe radius and stellar radius. During steady-state Roche-lobe overflow the stellar radius R_* and Roche lobe radius R_2 move in step,

$$\frac{\dot{R}_*}{R_*} = \frac{\dot{R}_2}{R_2}$$

15 Mass transfer is conservative if the total binary mass and angular momentum are conserved.

16 Mass transfer is unstable if the Roche lobe closes in on the star as a result of the transfer of a small amount of mass. This translates into an upper limit of order unity on the mass ratio of binaries with stable mass transfer.

17 Roche-lobe overflow occurs in the form of a geometrically thin, highly supersonic gas stream that follows a ballistic trajectory towards the accretor. The stream self-intersects and forms a plasma torus at the circularization radius.

3 STICKY STUFF: VISCOSITY AND ITS CAUSES

Accretion works only if the matter to be accreted successfully sheds excess angular momentum so it can spiral inwards. In this section we discuss a physical mechanism that achieves exactly that. The key behind the mechanism is that particles in the gas stream interact. They undergo close or distant collisions, thereby exchanging energy *and* momentum. Interactions on scales that are small compared to the radial and vertical extent of the disc, or even on microscopic scales, can nevertheless lead to an efficient transport of energy and linear or angular momentum over macroscopic scales. One such **transport phenomenon** is **viscosity**.

Viscosity is an effect that you are very familiar with in everyday life. Think of a cup of tea. You can easily stir the tea with a spoon, there is very little resistance to the motion of the spoon. If you try the same in a cup full of honey it takes a lot more effort to move the spoon. The honey is 'sticky' – it has a much higher viscosity.

3.1 Stress, strain and viscosity

To quantify viscosity, the degree of 'stickiness', we need to consider small elements of the fluid (tea, honey or stellar plasma) and how linear or angular momentum changes across the flow. Not surprisingly, this will involve *derivatives* of quantities that describe the plasma flow. Depending on the direction of the momentum transport relative to the flow direction we distinguish various types of viscosity. The **bulk viscosity** refers to the case where the flow velocity varies *along* the flow (Figure 34a). When the flow velocity varies *orthogonal* to the flow – in the presence of a shear flow (see, for example, Figure 34b) – there is an analogous shear viscous effect.

An accretion disc, where matter swirls around the central star with supersonic speed, represents a *rotating* fluid. The angular speed Ω of disc plasma at a distance R from the central accretor with mass M is almost always given by the Keplerian value

$$\Omega = \left(\frac{GM}{R^3}\right)^{1/2} \tag{29}$$

(see Block 1, Section 4.3.2). Note that we follow FKR and use Ω for the angular speed of accretion disc plasma, while previously we used ω to denote angular speed in general (see also the box entitled 'A coordinate system for accretion discs').

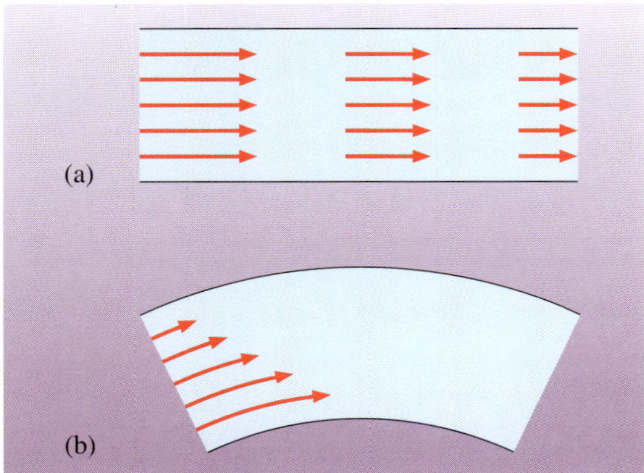

Figure 34 Velocity varying (a) along the flow and (b) orthogonal to the flow.

A COORDINATE SYSTEM FOR ACCRETION DISCS

In this block we describe the accretion disc in a frame of reference centred on the accretor, using cylindrical coordinates (R, ϕ, z), as shown in Figure 35. Note in particular the convention to use a capital R to denote the radial coordinate in this frame. This is the distance of a point *from the rotation axis*. The distance *from the primary star* is usually denoted by r; we have $r = R$ only in the disc mid-plane. The coordinate ϕ is the azimuth angle, the coordinate z the height above the disc mid-plane. A capital Ω denotes the angular speed of disc plasma orbiting the primary star.

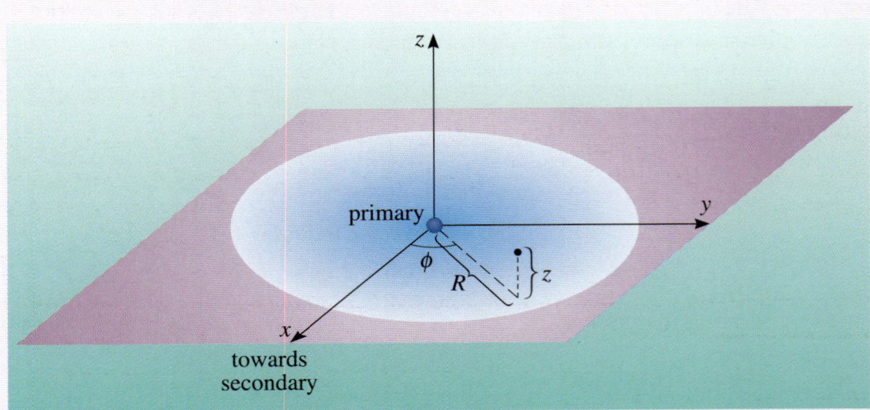

Figure 35 Accretion disc and cylindrical coordinates.

The disc flow is in fact a shear flow, and hence we have to consider the **shear viscosity** in some detail.

■ Why is the flow in an accretion disc a shear flow?

□ The gas in accretion discs essentially follows Keplerian orbits with local angular velocity $\Omega(R) \propto R^{-3/2}$ at distance R from the accretor. Consider two adjacent gas rings in the disc. As Ω drops with increasing R the inner ring slowly slides past (overtakes) the outer ring. ■

More generally, if the angular velocity Ω varies with distance R from the rotational axis we speak of **differential rotation**, as opposed to the more familiar solid-body rotation where the angular velocity is the same everywhere. In the case of a differentially rotating body, those parts of the body that are far away from the rotational axis may take longer (or less time) to complete one revolution than the parts that are closer to the axis. The body becomes distorted as a result of this differential rotation.

Question 23

Think of examples of differential rotation in everyday life. ■

The displacement of ‘blobs’ of gas in the presence of **turbulence** (i.e. random, large-scale motions of gas) can result in a high viscosity. We shall describe the viscosity operating in accretion discs mainly in terms of such a turbulent viscosity.

'STRESS CAUSES STRAIN'

This all-too-human rule has a more sober analogue in physics. It is most commonly found in the description of the deformation of a solid body under an external force applied to this body, e.g. the squeezing of a rubber ball. **Stress** measures the *force* applied over the surface of the body, while **strain** is a measure of the *deformation* the stress is causing. In the case of a fluid flow – such as our accreting stellar plasma – the 'deformation' manifests itself as a change in the velocity field in the flow.

More formally, stress denotes the force exerted per unit area on a surface. As the force F can have different directions with respect to the orientation of the area A there are different types of stress F/A (sometimes these different types are called stress components). For our purposes, the most important types are **pressure**, usually denoted by P, and **shear stress**. These are illustrated in Figure 36 for a small, cubic volume of the fluid flow. In the case of pressure the force is always perpendicular to the surface. In the case of shear stress the force is applied parallel to the surface. Imagine a fluid between two parallel plates, one of them stationary, the other moving with velocity v parallel to the other plate. A viscous fluid would develop a velocity profile like the one shown in Figure 36. The shear stress is the force needed to pull the plate of unit area with constant velocity v against the resistance of the fluid. A measure of the corresponding shear strain – the 'deformation' – is the gradient $\partial v/\partial z \approx v/l$ of the velocity in the shear flow perpendicular to the direction of motion of the plate.

In the simplest case the resulting strain is proportional to the applied stress. If we present this the other way round – the stress is proportional to the strain – then the viscosity is just the constant of proportionality:

stress = viscosity × strain

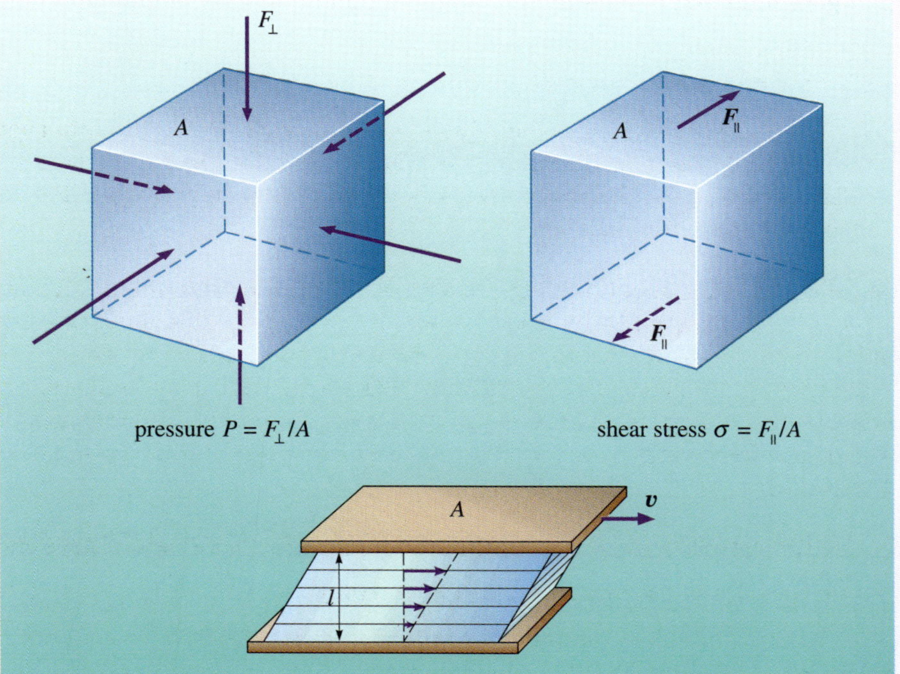

Figure 36 The definition of pressure and shear stress, and the velocity profile of a viscous fluid between two moving plates.

σ is the Greek letter sigma.

[handwritten notes in left margin:]
Shear stress ∝ strain
shear stress, $\sigma = \dfrac{F_{\parallel}}{A}$
strain $= \dfrac{\partial v}{\partial z}$
stress ∝ strain
$\sigma = \dfrac{F_{\parallel}}{A} = -\eta \dfrac{\partial v}{\partial z}$

In a fluid with large viscosity η, the shear stress σ for a given shear strain $\partial v/\partial z$ is large (see the box 'Stress causes strain'). Formally, this can be expressed as

$$\sigma = \frac{F_{\parallel}}{A} = -\eta \frac{\partial v}{\partial z} \tag{30}$$

Here η is also called **dynamical viscosity**. The force F_{\parallel} is along the flow with velocity v, while the velocity gradient is perpendicular to the flow (as in Figure 36).

In case you are worried about the symbol η: yes, we used the same symbol to denote the accretion efficiency in Section 1. Here it stands for the dynamical viscosity, and the two quantities have not the slightest thing in common. This is the first of several occasions in this block where a symbol (here a Greek letter) has multiple meanings. In most cases we are just following convention – as is FKR. It is quite likely that you will indeed encounter the different use of these symbols when you read other journal articles or books on a relevant subject. But relax: the context will always make the meaning clear.

v is the Greek letter nu.

[handwritten notes in left margin:]
kinematic viscosity is related to dynamic viscosity via density of the material
kinematic viscosity $\nu = \dfrac{\eta}{\rho}$
∴ $\eta = \nu\rho$
dynamic viscosity $\eta = \dfrac{\sigma}{\frac{\partial v}{\partial z}}$
2 diff. definitions for the same quantity
kinematic viscosity measures the viscosity relative to the density $\left(\dfrac{\eta}{\rho}\right)$

Confusingly, there is also a quantity, ν, called the **kinematic viscosity**, which is defined as

$$\nu = \eta/\rho \tag{31}$$

Question 24

What is the difference between dynamical and kinematic viscosity? ∎

3.2 Viscous torque and dissipation

A major result that is derived in a part of FKR which you do not read is that the kinematic viscosity due to the motion of particles or gas blobs with a characteristic speed v over a characteristic length λ is just

$$\nu \approx \lambda \times v \tag{32}$$

[handwritten annotation under equation:] ~ length × speed of particles

λ is the Greek letter lambda.

[handwritten notes in left margin:]
(25) $\nu = v\lambda$ hence kinematic viscosity must have units of $ms^{-1} \times m = m^2 s^{-1}$
$\eta = \rho\nu$ the dynamic viscosity has units of $kg\,m^{-3} \times m^2 s^{-1} = kg\,m^{-1}s^{-1}$
eq. 30 $F = \sigma = -\eta \dfrac{\partial v}{\partial z}$ $kg\,s^{-1}m^{-1} \times m\,s^{-1}\,m^{-1}$
$\dfrac{F}{A} = (kg\,m\,s^{-2}\,m^{-2}) = kg\,m^{-1}s^{-2}$ $= kg\,m^{-1}s^{-2}$

Activity 16
FKR section 4.6 (selected parts)

Question 25

What units does the kinematic viscosity ν have? Check if this is consistent with the definition of the dynamical viscosity in Equation 30. ∎

Activity 16 (45 minutes)

Viscous torques

You should now read selected parts of Section 4.6 in FKR.

Start by reading from the middle of the second paragraph on page 65, 'With the application to accretion discs in mind, …', then skip everything after 'Here the main pitfall appears' at the top of page 66. On page 67, read only the third paragraph (that starts with 'Because the chaotic motion …'). Resume reading the text at the bottom of page 67 ('For the case of circular rings of gas …') until the end of the section.

You might find this reading particularly hard going. Just keep on reading – even if you feel you do not really understand the arguments – the sequence of examples below will take you through the chain of arguments. In particular, by skipping over parts of Section 4.6 in FKR you might have difficulties in understanding how FKR Equation 4.27 comes about. This is the subject of one of the examples below, so just

[handwritten note at bottom left:] eq 4.27 $G(R) = 2\pi R \nu \Sigma R^2 \Omega'$

accept it as written for now. Also, do not worry about the details given after FKR Equation 4.28.

If you wish, you can revisit this reading once you have finished Section 3.2.

Keywords: **viscous torque**, **viscous dissipation**

Example 9

On page 67, FKR estimates the local mass exchange rate between adjacent gas rings as '$H\rho v$ per unit arc length'. Derive this expression.

Solution

We are considering a gas ring with radius R, height H and mass density ρ. The contact surface area between neighbouring gas rings is $2\pi RH$ (see Figure 37). If the typical speed of the chaotic motion of gas blobs is v, then a blob travels the radial distance $v\Delta t$ in a time Δt. In other words, all blobs in a volume $V = 2\pi RHv\Delta t$ cross the contact surface towards the centre of the disc. (On average, an equal number of blobs cross the surface towards the outer edge of the disc.) The mass m_{blob} of the blobs crossing the surface is just $m_{\text{blob}} = V\rho$, so the rate of mass crossing is $\dot{m}_{\text{blob}} = V\rho/\Delta t$, and the rate of crossing per unit arc length is $\dot{m}_{\text{blob}}$, divided by $2\pi R$. This simplifies to

$$\frac{V\rho}{\Delta t}\frac{1}{2\pi R} = \frac{2\pi RHv\rho\Delta t}{\Delta t}\frac{1}{2\pi R} = H\rho v$$

as claimed.

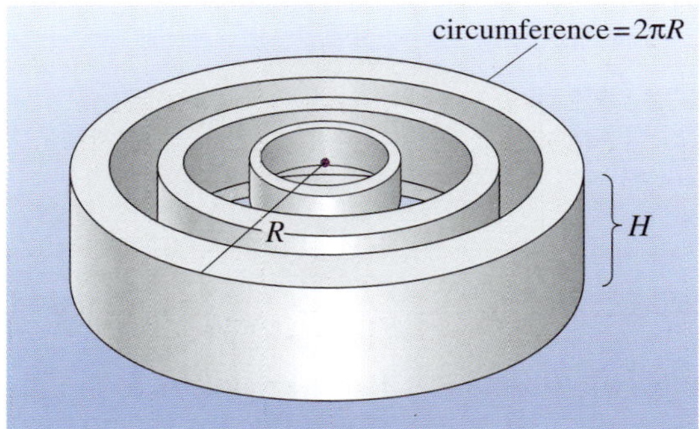

Figure 37 Accretion disc rings.

A rather useful concept in the study of accretion discs is the use of vertically integrated quantities. We make clear why and what these are later in the block, but at this point we make use of the most prominent of the integrated quantities: the **surface density** Σ (of mass).

Σ is the Greek capital letter sigma. Note that here Σ is *not* used to denote summation.

Formally, Σ is defined by

$$\Sigma(R) = \int_{-\infty}^{+\infty} \rho(R,z)\,\mathrm{d}z \qquad (33)$$

This integrates the gas density ρ in a direction perpendicular to the disc from very far below ($z \to -\infty$) to very far above ($z \to \infty$) the disc mid-plane. The third coordinate, the azimuth ϕ, does not appear explicitly because the disc is axisymmetric, so that neither ρ nor Σ depend on ϕ. The surface density tells us how much mass a disc ring with unit surface area at radius R contains.

■ What are the SI and cgs units of Σ?

❑ As Σ is mass per surface area the unit must be kg m^{-2} (or g cm^{-2}). Another way of seeing this is by recognizing that integration over z contributes a length (m), and ρ obviously has the unit kg m^{-3}. Hence m $\times$ kg m^{-3} = kg m^{-2}. ■

It is useful to relate the surface density Σ to the familiar volume mass density ρ in a simple way. If ρ is constant in z-direction, and the full disc thickness is just H, then Equation 33 reduces to

$$\Sigma = H\rho \tag{34}$$

Even if the density varies with height z it is still possible to write down an equation like Equation 34. In that case H is a characteristic vertical scale height of the disc, and ρ a typical density in the disc at radius R. FKR makes abundant use of this so-called *one-zone model*. If you have trouble understanding what this means, just picture the accretion disc as a disc with finite thickness and constant density at each radius, as in Figure 37. We come back to the detailed vertical structure of accretion discs a little later in this block.

In the following example we are considering the viscous torque in the accretion disc.

Example 10

Note that in Block 1 the symbol Γ has been used to denote torque. Note also that elsewhere in Block 3 the letter G may denote the gravitational constant.

Consider two adjacent gas rings in an accretion disc at radius R. The torque exerted by the outer ring on the inner ring is $G = 2\pi R\nu\Sigma R^2\Omega'$ (FKR Equation 4.27). Derive this expression, starting from the definition of torque and viscous force. (The symbol Ω' ('omega prime') stands for the derivative of Ω with respect to R, i.e. $\Omega' = \mathrm{d}\Omega/\mathrm{d}R$).

Solution

The magnitude of torque is force times lever arm (see, for example, Block 1, Section 4.5). The force in our case is a viscous force, hence can be written as F = contact surface area $\times$ viscous stress = $A \times \sigma$. The contact surface area of the two disc rings is $A = 2\pi RH$ (see Figure 37). To calculate the viscous stress we consider our Equation 30, which defines the dynamical viscosity. We need to apply this equation in the context of a rotating fluid. As the orbital speed of gas in the ring is just $v = R\Omega$, we might simply replace the gradient of the velocity v with the product $R\mathrm{d}\Omega/\mathrm{d}R$ which involves the gradient of the angular speed. Doing this the shear stress becomes

$$\sigma = -\eta R \frac{\mathrm{d}\Omega}{\mathrm{d}R} \tag{35}$$

where η is the dynamical viscosity. You might think that perhaps $\mathrm{d}(R\Omega)/\mathrm{d}R$ should be used in Equation 35 instead of $R\,\mathrm{d}\Omega/\mathrm{d}R$. It is easy to see why not. Using the product rule tells us that

$$\mathrm{d}(R\Omega)/\mathrm{d}R = R\,\mathrm{d}\Omega/\mathrm{d}R + \Omega\,\mathrm{d}R/\mathrm{d}R = R\,\mathrm{d}\Omega/\mathrm{d}R + \Omega$$

and this is non-zero (i.e. equal to Ω) even in the *absence* of shear. But viscous stresses exist only in the *presence* of shearing motion. If Ω = constant there is no shear, hence no stress, i.e. we *must* have $\sigma = 0$ in this case. Therefore only the first term, $R\,\mathrm{d}\Omega/\mathrm{d}R$, can contribute to the shear stress σ.

Now using the expressions for A and σ, and observing that $\eta = \rho\nu$ and $\Sigma = H\rho$ we find for the magnitude of the torque

$$G = FR = -A\sigma R = 2\pi RH \times \eta R \frac{\mathrm{d}\Omega}{\mathrm{d}R} \times R = 2\pi RH \times \rho\nu R \frac{\mathrm{d}\Omega}{\mathrm{d}R} \times R$$

hence $G = 2\pi R\nu\Sigma R^2\,\mathrm{d}\Omega/\mathrm{d}R$. Note the sign of G: we introduced a minus sign in the above, which cancelled with the minus sign of Equation 35. Although we said we calculate the magnitude of the torque, the value we obtain is still either positive or negative. The significance of this is as follows: implicit in our above derivation was the assumption that the torque vector is parallel or antiparallel to the rotational axis of the accretion disc. If it is parallel and in the same direction as the vector $\boldsymbol{\Omega}$, G is a positive quantity, and the torque acts in the same direction as the fluid rotates. In other words, the rotating fluid speeds up. Conversely, if the torque is antiparallel to the vector $\boldsymbol{\Omega}$, the quantity G is negative, and the torque brakes the rotation. In the case of a Keplerian accretion disc we have $\Omega(R) \propto R^{-3/2}$ and hence $\mathrm{d}\Omega/\mathrm{d}R < 0$ and also $G < 0$. This shows that G is the torque exerted by the slower outer ring on the faster inner ring: the slower ring attempts to hold back the faster inner ring as it glides past it. The torque exerted by the inner ring on the outer ring is just $-G$. ■

A word of clarification on derivatives might be in order. In our cylindrical coordinate system a quantity like Ω could in principle depend on all three coordinates R, z and ϕ. This dependence can be characterized by the three partial derivatives, $\partial\Omega/\partial R$, $\partial\Omega/\partial z$ and $\partial\Omega/\partial\phi$, i.e. derivatives along the three coordinate axes. The partial derivative $\partial\Omega/\partial R$ expresses the local change of Ω when we march in the direction of R, i.e. along a line with z = constant and ϕ = constant. In the framework of the one-zone model Ω does not depend on z. As the disc is also axisymmetric, Ω does not depend on ϕ either. Therefore, in our case Ω depends *only* on R. Hence there is no difference between the partial derivative $\partial\Omega/\partial R$ and the standard notation $\mathrm{d}\Omega/\mathrm{d}R$, and indeed you might find both used interchangeably.

The expression you derived in Example 10 is in fact one of the major equations we will need again later in the block. Therefore we repeat it here.

The viscous torque is given by FKR Equation 4.27 (where $\Omega' = \partial\Omega/\partial R$)

$$G = 2\pi R\nu\Sigma R^2 \frac{\partial\Omega}{\partial R} \tag{36}$$

Question 26

Verify that the right-hand side of Equation 36 has the units of a torque. ■

Did you find the paragraph of FKR containing FKR Equations 4.28–4.30 a little confusing? Let us reiterate what FKR is saying. First, the important information to take away is the expression that describes the amount of heat generated in the plasma flow by friction, per unit time and unit area:

$$D(R) = \frac{1}{2}\nu\Sigma\left(R\frac{\partial\Omega}{\partial R}\right)^2 \qquad \text{(FKR 4.29) (37)}$$

The quantity $D(R)$ is the rate, per unit surface area, at which the mechanical energy of the rotational motion of the plasma is converted into heat due to viscosity. It is called the **viscous dissipation** rate, for short.

[handwritten left margin:]
(27)

dissipation rate measures conversion of E per unit of time per surface area. E is force × distance so unit is kg m² s⁻². and here unit of dissipation rate is (kg m² s⁻²) s⁻¹ m⁻² = kg s⁻¹

$$D(R) = \frac{1}{2} \nu \Sigma \left(R \frac{d\Omega}{dR} \right)^2 \quad \text{kg s}^{-1}$$

eq. 29 $\Omega = \left(\dfrac{GM}{R^3} \right)^{\frac{1}{2}}$ angular speed of plasma

[handwritten above Question 27:] rate per unit surface area at which mechanical E of rotation motion of plasma is converted into heat due to viscosity

Question 27

What is the SI unit of the viscous dissipation rate? ■

You can see why this dissipation must occur as follows. Consider three adjacent gas rings which rotate with their local Keplerian angular speed (Equation 29). We label the rings as A, B and C, starting from the inner ring (Figure 38). The inner edge of ring B is at radius R, the outer edge at radius $R + \Delta R$. A rotates faster than B and hence tries to spin-up (increase the speed of) ring B, i.e. there is a positive torque $G(R)$ on B from A. Conversely, ring C rotates slower than B and tries to spin-down B, i.e. there is a negative torque $-G(R + \Delta R)$ on B from C. Because of the radial dependence of the viscous torque G, the sum of these two torques is non-zero. In other words, there is a net torque on ring B (in this case trying to increase its spin). This involves work, or, to be precise, a rate of working, i.e. power. To find an expression for the rate of working think of an analogy involving linear motion. If you pull a sledge horizontally in a straight line you exert a constant pulling force of magnitude F to overcome friction with the ground and maintain a constant speed v. The pulling power is $dW/dt = Fv$ (where W denotes work). Likewise, in the case of rotational motion the rate of working is given by $dW/dt = G\Omega$. The magnitude of torque G takes the place of the magnitude of the force F, the angular speed Ω the place of the translational speed v.

[handwritten left margin:]
rate of work = $\dfrac{dW}{dt} = G\Omega$

a fraction of $\frac{dW}{dt}$ deposits mechanical energy into disc in form of heat

This viscous dissipation deposits E into ring at rate $G(R) \times \Delta\Omega$ ($\Delta\Omega$ is difference in angular speed at outer + inner edge of gas ring) NOTE

Some fraction of the rate of working that is associated with the net torque deposits mechanical energy into the disc in the form of heat. In fact, it turns out that this viscous dissipation deposits energy into the ring at a rate $G(R) \times \Delta\Omega$, where $\Delta\Omega$ is the *difference* between the angular speed at the outer and inner edge of the gas ring. (This reasoning is sufficient for the purposes of the course; a more thorough

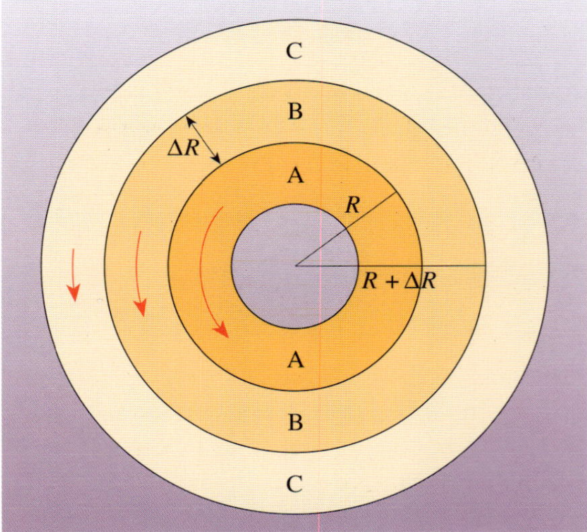

Figure 38 Torque balance for an accretion disc ring.

justification is hinted at in FKR.) Making use of the first-order expansion (see Block 1, Section 3.7, and Example 12 below) the difference $\Delta\Omega$ is approximately

$$\Delta\Omega \approx \frac{\mathrm{d}\Omega}{\mathrm{d}R}\Delta R$$

if the width ΔR of the ring is small. So viscous dissipation occurs at a rate $G(R) \times \Delta\Omega = G \times \Delta R(\mathrm{d}\Omega/\mathrm{d}R)$. We normalize this now to the surface area of the disc ring. The upper and lower side of the disc ring each has an area of $2\pi R\Delta R$ ('circumference × width'; this is also the subject of Examples 11 and 12). Hence the dissipation rate per unit surface area is

$$D = \frac{G\,\Delta R(\mathrm{d}\Omega/\mathrm{d}R)}{2 \times 2\pi R\Delta R} = G\frac{\mathrm{d}\Omega/\mathrm{d}R}{4\pi R}$$

Using Equation 36 this becomes

$$D = 2\pi R\nu\Sigma R^2\frac{\mathrm{d}\Omega}{\mathrm{d}R} \times \frac{\mathrm{d}\Omega/\mathrm{d}R}{4\pi R}$$

This simplifies to Equation 37.

Example 11

Show that the surface area of a narrow disc annulus at radius R and with width ΔR is $2\pi R\Delta R$.

Solution

To work out the area properly we could calculate the difference between the surface areas of two circles with radius $R + \Delta R$ and radius R:

$$\text{area} = \pi(R + \Delta R)^2 - \pi R^2 = \pi[R^2 + 2R\Delta R + (\Delta R)^2] - \pi R^2$$

The first and last term cancel, leaving

$$\text{area} = 2\pi R\Delta R + \pi(\Delta R)^2 = 2\pi R\Delta R \times \left(1 + \frac{\Delta R}{2R}\right)$$

Now, since the ring is narrow we have $\Delta R \ll R$ and thus $\Delta R/R \ll 1$. So the second term in brackets is very small compared to the first one and can be neglected. Formally, we say that *to first order (in* ΔR*)* the surface area of the disc ring is $2\pi R\Delta R$, i.e. 'circumference × width'. This is not at all surprising, since you could have cut the ring radially at one point and straightened it into a long, thin rectangle, the area of which is, of course, length × width. The small difference between the two long sides of the straightened rectangle is negligible if the width is very small. Henceforth we will make use of such approximations, or similar ones, by just appealing to intuition. When you come across an approximation you do not trust, though, you should try to derive the conditions under which it holds rigorously, as above. ■

Before moving on, let's use the disc annulus surface area to convey the essence of another rather powerful technique: first-order expansion.

Example 12

Use the first-order expansion of the area of a circle to derive the expression $2\pi R\Delta R$ for the surface area of a disc annulus (see Block 1, Section 3.7).

Solution

The Taylor expansion of an arbitrary function $f(R)$ of the variable R in the vicinity ΔR of a specific value of R is up to first order

$$f(R+\Delta R) \approx f(R) + [(R+\Delta R) - R]\left.\frac{\mathrm{d}f}{\mathrm{d}R}\right|_R = f(R) + \Delta R\frac{\mathrm{d}f}{\mathrm{d}R}$$

Hence the difference Δf between the value of f at R and $R + \Delta R$ is

$$\Delta f = f(R+\Delta R) - f(R) \approx \Delta R\frac{\mathrm{d}f}{\mathrm{d}R}$$

This *first-order expansion*

$$\Delta f \approx \Delta R\frac{\mathrm{d}f}{\mathrm{d}R} \qquad (38)$$

is useful in many circumstances, for any quantity f. In fact, we used it for the estimate $\Delta\Omega \approx \Delta R(\mathrm{d}\Omega/\mathrm{d}R)$ above.

We now use it for the area $A(R)$ of a circle with radius R, where $f(R) = A(R) = \pi R^2$. As

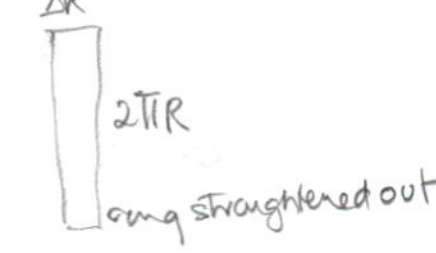

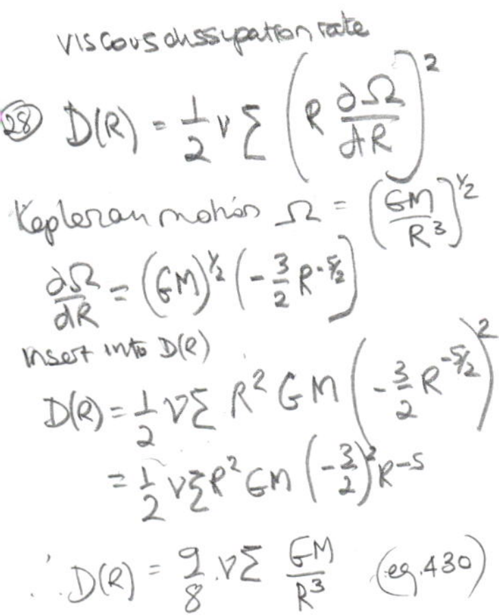

$$\frac{\mathrm{d}A}{\mathrm{d}R} = \frac{\mathrm{d}(\pi R^2)}{\mathrm{d}R} = 2\pi R$$

we have indeed $\Delta A = 2\pi R\Delta R$ for the area of the disc ring. ■

Question 28

Using the angular velocity for Keplerian motion (Equation 29), derive FKR Equation 4.30 from Equation 37. ■

3.3 The cause of viscosity

We now turn to the physical origin of viscosity in accretion discs.

One important form of viscosity that is always present in a plasma is the molecular viscosity. It comes about due to the motion of ions in the plasma. A plasma particle (an ion or electron) travels a typical distance, the **deflection length** λ_d, before it is appreciably deflected from its initial trajectory, due to the effect of many weak, 'distant' collisions with other plasma particles. The deflection length can be calculated if the plasma temperature and density are known. In other words, the ions travel for roughly one deflection length λ_d before their linear momentum changes. The ions have thermal speeds of order the sound speed c_s, hence according to Equation 32 there is a viscosity of order $\nu \approx \lambda_d c_s$, the **molecular viscosity**.

It turns out that the molecular viscosity is far too weak, by many orders of magnitude, to bring about the dissipation and angular momentum transport required in accretion discs. (We will learn more about how observed phenomena tell us about the magnitude of viscosity later in the block.) But if it is not the molecular viscosity, what else can it be?

REYNOLDS NUMBER

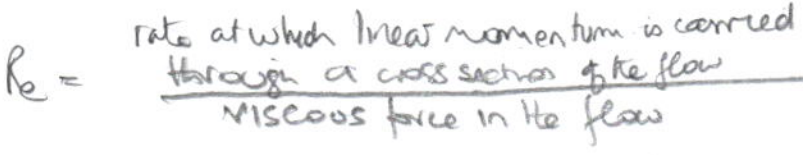

The Reynolds number Re of a fluid flow is a dimensionless parameter characterizing the relative importance of the viscous force for the dynamics of the flow. More precisely, Re is the ratio of the rate at which linear momentum is carried through a cross-section of the flow, to the viscous force in the flow. If $Re \ll 1$ viscous forces dominate the flow, while if $Re \gg 1$ viscous effects are dynamically unimportant.

Laboratory experiments show that if the Reynolds number of a fluid flow is gradually increased, turbulence sets in once the Reynolds number is above a critical limit.

The **Reynolds number** of the flow provides a clue. In a typical Keplerian accretion disc the Reynolds number due to the action of molecular viscosity alone is very large, $>10^{14}$.

Activity 17

The α-viscosity

Now continue with FKR and read Section 4.7. Start with the second paragraph of page 70 where it says 'Curiously, a clue to the right kind of viscosity is …'.

The α-prescription of viscosity introduced in this reading is known as **α-viscosity**.

Keywords: **turbulence**, **turbulent eddy**, **turnover velocity**, **α-prescription** ■

■ In the functional form $\nu = \upsilon\lambda$ for the kinematic viscosity, what do υ and λ represent in the case of turbulent viscosity?

❑ υ is the typical speed of the largest turbulent eddies (blobs of gas), and λ is a typical size of the largest turbulent eddies. ■

Question 29

Explain, in your own words, the meaning of 'α-viscosity'. ■

As the concept of the α-viscosity turns out to be very useful – and is in fact very widely used – we repeat the definition here

$$\nu = \alpha c_s H \qquad \text{(FKR 4.34) (39)}$$

This relation is purely *phenomenological*, i.e. it expresses the fact that, on dimensional grounds, it is reasonable to assume that the viscosity scales as the product of c_s and H. It is *not* a physical law or the result of a strict derivation that has started from physical laws.

Ever since the phenomenological α-viscosity was introduced astrophysicists have tried to identify a proper physical mechanism that generates a viscosity of the right magnitude and sign. A currently very promising candidate is magneto-hydrodynamic

(MHD) turbulence. This involves weak magnetic fields in the accretion disc plasma. The details of MHD turbulence are beyond the scope of this course; all we can say here is that the interaction of the shearing plasma flow and the initially weak magnetic field leads to an amplification of the field, and to turbulent motion in the disc. The researchers Steven Balbus and John Hawley rediscovered the importance of this instability for accretion discs. MHD turbulence is an area of intense and active research. By the time you read this sentence it is likely that yet newer discoveries have been made. You are welcome to browse Section 4.8 of FKR for a feeling of the issues involved, but you might find some of the material presented there too condensed for easy digestion. We do not offer to help you here, as this is voluntary reading and not required for the course.

The name game: part 3

Once again, we suggest you skim-read through Section 3, and make a list of all symbols that denote *new* variables and *new* quantities, and note down what they mean. Flag up those symbols that have (unfortunately) multiple uses.

Once you have compiled your list and explained the meaning of all entries, compare it to the one given in Appendix A5 at the end of the Study Guide.

3.4 Summary of Section 3

1. Viscosity is a transport phenomenon and describes the relation between stress and strain, stress = viscosity × strain.
2. Stress is the force exerted per unit area on a surface. Strain is a measure for the deformation the stress is causing.
3. The dynamical viscosity η is the constant of proportionality between strain and stress. For fluid flows, the most important types of stress are pressure (force perpendicular to the surface) and shear stress (force applied parallel to the surface). The viscous shear stress is
$$\sigma = -\eta R \frac{\mathrm{d}\Omega}{\mathrm{d}R}$$
4. The kinematic viscosity ν is the dynamical viscosity η divided by the density ρ of the medium, $\nu = \eta/\rho$.
5. The kinematic viscosity due to the motion of particles or gas blobs with a characteristic speed υ over a characteristic length λ is $\nu \approx \lambda \times \upsilon$.
6. Ions in a plasma have thermal speeds of order of the sound speed c_s and travel for roughly one deflection length λ_d before their linear momentum changes. The corresponding viscosity $\nu \approx \lambda_\mathrm{d} c_\mathrm{s}$ is called the molecular viscosity. In the case of turbulent viscosity υ is the typical speed and λ the typical size of the largest turbulent eddies.
7. Accretion discs are best described in a cylindrical coordinate system (R, ϕ, z) with the origin at the centre of the accretor.
8. Vertically integrated quantities can be used to describe the radial disc structure. One of them is the surface density Σ
$$\Sigma(R) = \int_{-\infty}^{+\infty} \rho(R, z)\,\mathrm{d}z$$

9 In the one-zone model the surface density is $\Sigma = H\rho$, where H is the vertical disc scale height.

10 The magnitude of the viscous torque in an accretion disc at radius R with local angular speed Ω is

$$G = 2\pi R\nu\Sigma R^2 \frac{\partial\Omega}{\partial R}$$

11 Viscous dissipation converts mechanical energy of the rotational motion of the disc plasma into heat. The conversion rate per unit surface area is

$$D(R) = \frac{1}{2}\nu\Sigma\left(R\frac{\partial\Omega}{\partial R}\right)^2$$

12 The Reynolds number Re of a fluid flow is the ratio of the inertia in the flow (the rate at which linear momentum is carried through a cross-section of the flow) to the viscous force in the flow. In a Keplerian accretion disc the Reynolds number due to the action of molecular viscosity is very large ($>10^{14}$). Laboratory experiments suggest that turbulence is likely in flows with large Reynolds number.

13 The phenomenological α-viscosity expresses the disc viscosity in terms of the vertical disc scale height H and the sound speed c_s: $\nu = \alpha c_s H$.

14 A currently very promising candidate for the physical mechanism that provides the viscosity in accretion discs is magneto-hydrodynamic turbulence.

4 DISCS AT PEACE: STEADY-STATE ACCRETION

Using the background of the physical principles introduced in the preceding sections, and of those reviewed in Block 1, we shall now develop a quantitative model for accretion discs. The purpose of a physical model is to check our understanding of the physics governing the system we are trying to describe. We shall use the model to predict certain properties of the accretion process in a close binary star, e.g. the temperature of the disc, and test these predictions against observations or experiment. Usually such a reality check will quickly reveal the general usefulness of the model. The feedback from experiment usually allows one to refine and extend the model, and thus to improve on the understanding of the underlying physics. This way we may uncover a piece of physics that was missing from the original model, but now proves vital to make it work. Sometimes such a discovery has consequences far beyond the immediate context of the model – here the plasma flow in a close binary system – and thus contributes another piece to the puzzle of the physical world we live in. This is research at work!

As always in physics most insight is gained when the model is set up in as simple a way as possible. That is to say, a good model should capture the essential physical ingredients of the process, but not dwell on unnecessary details. Once the essentials are understood, more details can be added. The art of being a good physicist is to cut through the undergrowth – select only the important ingredients. Einstein once said: physics should be made as simple as possible, but not simpler.

In this spirit, we set out with a number of simplifying assumptions for accretion discs. As detailed below we assume that the discs really are what the name suggests – flat. We shall see that the vertical and radial structure of the disc can be treated separately; they are decoupled. We furthermore assume that there are no external magnetic fields that could interfere with the accretion flow, an assumption we shall relax in Section 7. The most important assumption will be introduced in Section 4.2, where we consider steady-state discs. 'Steady state' means that the accretion flow pattern is the same whenever we look. In technical terms: the quantities in the disc equations we are going to develop will have no explicit dependence on time. We shall relax the steady-state assumption in Section 8. There will be more on steady-state discs in Section 5.

4.1 Conservation laws

The first assumption to make is that the accretion disc is *geometrically thin*. We assume that the disc is effectively confined to the orbital plane, i.e. the plasma density falls off rapidly when we move away from the plane. Accretion discs are, in fact, rather like flat stars, an analogy we shall pursue further in Section 5.2. As you have already seen above, for an analysis of the *radial* structure of the disc it is practical to operate with quantities that are vertically integrated, or averaged over the vertical direction, a concept you first met in Section 3.2.

To consider the accretion flow in the disc plane a bit more quantitatively we rewrite two fundamental conservation laws of physics, the conservation of mass and of angular momentum, in terms of vertically integrated or averaged quantities (variables).

(so)

In particular, the following reading from FKR introduces three main equations of particular interest, all of them **partial differential equations**. We shall discuss them in detail after the reading. But we should clarify a few things in advance that may help you to navigate this rather challenging reading in FKR.

(80)

In general, the physical quantities describing the disc, such as the surface density Σ, the viscosity ν and the radial drift velocity v_R depend on both the time t and the radial distance R from the centre of the disc. These quantities are functions of two independent variables, R and t, e.g. $\Sigma = \Sigma(R, t)$. Equations involving these functions include partial derivatives with respect to these variables. Recall that the partial derivative $\partial\Sigma/\partial R$ denotes the rate of change of Σ with distance R for a fixed t, i.e. this is the radial surface density gradient in the disc at a given time (see also Figure 39).

- Describe the meaning of the partial derivative $\partial\Sigma/\partial t$.

NOTE

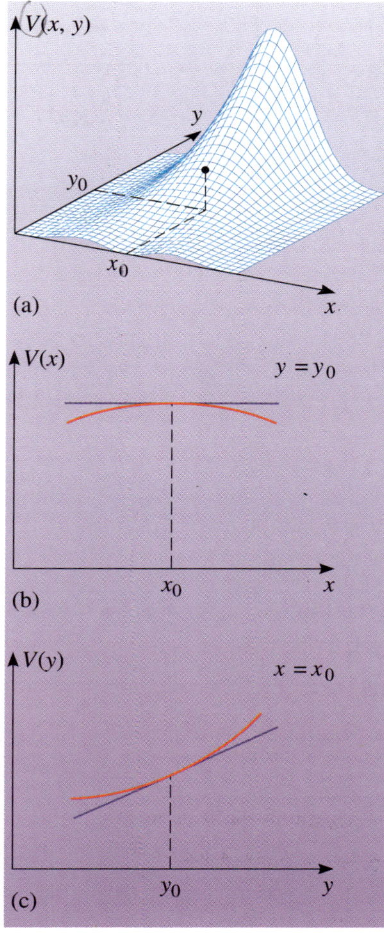

- ❑ This denotes the rate of change of Σ with time, at a fixed distance R. Therefore, this is the rate at which a disc ring at distance R gains or loses mass. ∎

NOTE In principle, the same comments apply to the angular speed Ω, except that the time-dependence is almost always assumed to vanish (i.e. $\partial\Omega/\partial t = 0$). As in FKR Equation 4.27, the derivative of Ω with respect to R is written as Ω', i.e. $\Omega' \equiv \partial\Omega/\partial R$. This derivative is, of course, itself a function of R.

And as a final complication – beware that the symbol G is used in FKR *both* for the NOTE gravitational constant *and* the viscous torque. The meaning should be clear from the context, though.

Activity 18 **(30 minutes)**

Radial disc structure

Read Section 5.1 and Section 5.2 of FKR up to the end of the first paragraph on page 82 ('… can be solved by separation of variables').

This involves a lot of algebra, as on page 81 of FKR. But there is no reason to be frightened. Below we provide plenty of hints to help you through the material. Just take your time and try to follow the arguments and manipulations step by step. If you have serious difficulties, you can also read the hints in parallel with the text in FKR.

Keywords: **thin disc approximation**, **drift velocity**, **radial drift velocity**, **diffusion equation** ∎

4.1.1 Conservation of mass

The first equation to note describes the conservation of mass (FKR Equation 5.3),

$$R\frac{\partial\Sigma}{\partial t} + \frac{\partial}{\partial R}(R\Sigma v_R) = 0 \qquad (40)$$

In deriving this equation FKR considers a disc ring (annulus) between radii R and $R + \Delta R$ (see Figure 40 overleaf). The disc ring is supposed to be rather narrow, i.e. we have $\Delta R \ll R$. The mass stored in the ring can be calculated as 'surface density × surface area'. From Example 11 or Example 12 we know that the surface area of the disc ring is just $2\pi R\Delta R$, i.e. 'circumference × width'. Hence the mass M_a in the disc annulus is $2\pi R\Delta R \times \Sigma$. Mass conservation implies that mass cannot appear out

(Surface area × density)

Figure 39 Diagram illustrating the meaning of partial derivatives. The quantity V is a function of both x and y.

Activity 18
 FKR Section 5.1, 5.2
 (5.2 only up to 1st paragraph)
 on p. 82

thin disc approx – simplifying assumption that the thickness H of an accretion disc at distance R from the accretor is small cpd to R. This allows one to treat the vertical + radial structure of disc separately. + motivates the use of vertically integrated quantities like surface density to describe the disc

drift velocity – the radial component of the velocity of matter in an accretion disc. The radial inward + outward drift due to viscous diffusion is v. slow cpd with Keplerian motion

v_R radial drift velocity – the velocity with which matter moves radially in an accretion disc.

diffusion equation . a differential equation where the derivative of a quantity is proportional to its 2nd derivative

$$\frac{dy}{dt} \propto \frac{d^2y}{dt^2}$$

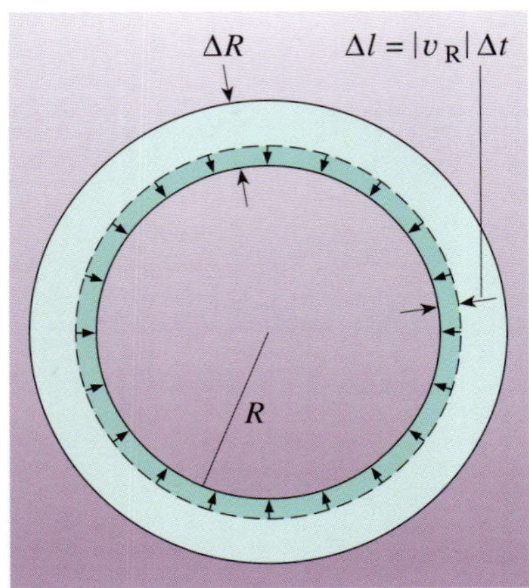

Figure 40 Mass crossing the inner boundary of the disc ring at radius R due to the radial drift with velocity v_R.

of nowhere. So when the mass in the disc ring increases it must be because more mass flows into the ring than out of it. In other words, the rate of change of the ring mass is given as the difference between the mass inflow and outflow rate.

The rate of change of the disc ring mass M_a is

$$\frac{\partial M_\mathrm{a}}{\partial t} = \frac{\partial(2\pi R \Delta R \Sigma)}{\partial t}$$

As 2π, ΔR and R do not depend on time this simplifies to

$$\frac{\partial M_\mathrm{a}}{\partial t} = 2\pi R \Delta R \frac{\partial \Sigma}{\partial t} \tag{41}$$

To calculate the net inflow/outflow of mass we first define another quite useful quantity. The local **mass accretion rate** $\dot{M}(R, t)$ is the amount of mass that flows per unit time through the boundary at radius R between adjacent disc annuli. By convention $\dot{M}$ is positive if the mass flows inwards, towards smaller radii, and negative if the mass flows to larger radii. We can relate $\dot{M}$ to the radial drift velocity v_R, which is defined to be positive if the material drifts outwards. Assume that the disc gas drifts inwards with velocity $(-v_\mathrm{R}) > 0$. Then in a short time interval Δt all the mass $\delta M < M_\mathrm{a}$ which is originally in a narrow sub-annulus of width $\Delta l = (-v_\mathrm{R})\Delta t < \Delta R$ will be able to cross the boundary surface at R (see Figure 40). Hence the local mass accretion rate is

$$\dot{M} = \frac{\delta M}{\Delta t} = \frac{2\pi R \Delta l \Sigma}{\Delta t} = 2\pi R \frac{\Delta l}{\Delta t} \Sigma = 2\pi R (-v_\mathrm{R}) \Sigma$$

Here we have calculated the mass in the sub-annulus in the same way as above for the full disc ring. Hence, we have

$$\dot{M}(R, t) = -2\pi R v_\mathrm{R} \Sigma \tag{42}$$

To work out the net change ΔM_a of the mass M_a of the disc ring between radii R and $R + \Delta R$ in the time interval Δt we simply have to take the difference between the local mass flow rate at $R + \Delta R$ and at R, and multiply it by Δt. Remembering the definition of partial derivatives and the approximation by first-order expansion (Example 12) we see that the net flow can be written as

$$\Delta M_\mathrm{a} = [\dot{M}(R + \Delta R, t) - \dot{M}(R, t)] \times \Delta t \approx \Delta R \frac{\partial \dot{M}}{\partial R} \times \Delta t$$

Using Equation 42 this becomes

$$\frac{\Delta M_\mathrm{a}}{\Delta t} = \Delta R \frac{\partial \dot{M}}{\partial R} = -\Delta R \frac{\partial(2\pi R v_\mathrm{R} \Sigma)}{\partial R}$$

i.e.

$$\frac{\Delta M_\mathrm{a}}{\Delta t} = -2\pi \Delta R \frac{\partial(v_\mathrm{R} R \Sigma)}{\partial R} \tag{43}$$

For small time intervals Δt the expression $\Delta M_\mathrm{a}/\Delta t$ in Equation 43 becomes the derivative $\partial M_\mathrm{a}/\partial t$. Equating the right-hand side of Equation 41 with the right-hand side of Equation 43, and dividing by $2\pi \Delta R$, finally reproduces Equation 40. The derivation is exact in the limit $\Delta R \to 0$.

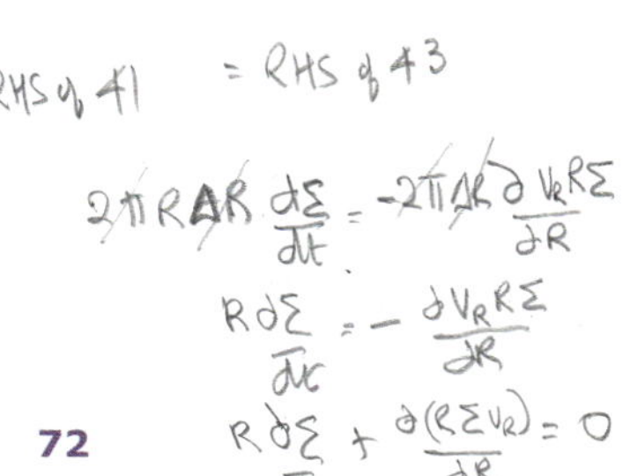

4.1.2 Conservation of angular momentum

The second equation to note,

$$R\frac{\partial}{\partial t}(\Sigma R^2\Omega) + \frac{\partial}{\partial R}(R\Sigma v_\mathrm{R} R^2\Omega) = \frac{1}{2\pi}\frac{\partial G}{\partial R} \qquad \text{(FKR 5.4) (44)}$$

expresses the conservation of angular momentum. As we shall see the derivation of this equation is analogous to the one for mass conservation, with the exception that an additional term occurs due to the presence of a torque.

■ Consider the first two terms in Equation 44, and compare them to the two terms in the mass conservation Equation 40. Do you see any similarity?

❑ The terms in Equation 44 are obtained by replacing Σ in Equation 40 with $\Sigma R^2\Omega$. ■

This is not a coincidence. As $R^2\Omega$ is the specific angular momentum of the disc material, i.e. the angular momentum per unit mass, $\Sigma R^2\Omega$ is the angular momentum of the disc material per unit surface area. The two terms represent the change of the angular moment of the disc ring due to an imbalance between incoming angular momentum that sticks to the mass flowing into the ring, and the outgoing angular momentum that sticks to the mass flowing out of the ring.

Question 30

Following the example of Section 4.1.1, derive the first two terms of Equation 44 explicitly. ■

The additional term $\frac{1}{2\pi}\frac{\partial G}{\partial R}$ on the right-hand side of Equation 44 arises because there is also a net viscous torque acting on the disc ring, as we have mentioned above in the context of viscous dissipation (Section 3.2). By analogy to Newton's second law – 'force equals rate of change of linear momentum' – the net torque contributes a rate of change of angular momentum. The torque term constitutes a source or sink of angular momentum. The net viscous torque G on the disc ring is just

$$G(R+\Delta R) - G(R) \approx \Delta R\frac{\partial G}{\partial R}$$

As Equation 44 is obtained from the angular momentum balance equation for the disc ring by division by $2\pi\,\Delta R$ (see the last step in Section 4.1.1) the final source term in Equation 44 is

$$\frac{1}{2\pi\Delta R}\Delta R\frac{\partial G}{\partial R} = \frac{1}{2\pi}\frac{\partial G}{\partial R}$$

as required.

4.1.3 Viscous diffusion

A clever manipulation of Equations 44 and 40 to eliminate the radial drift velocity (see Example 13 and Question 31 below) delivers us FKR Equation 5.8, probably the most important equation of this section. Using the form of G we derived earlier (also given in FKR Equation 5.5), and assuming that the disc rotates with the

Keplerian angular speed $\Omega(R) = (GM/R^3)^{1/2}$, gives

$$\frac{\partial \Sigma}{\partial t} = \frac{3}{R}\frac{\partial}{\partial R}\left[R^{1/2}\frac{\partial}{\partial R}(\nu \Sigma R^{1/2})\right] \qquad \text{(FKR 5.8) (45)}$$

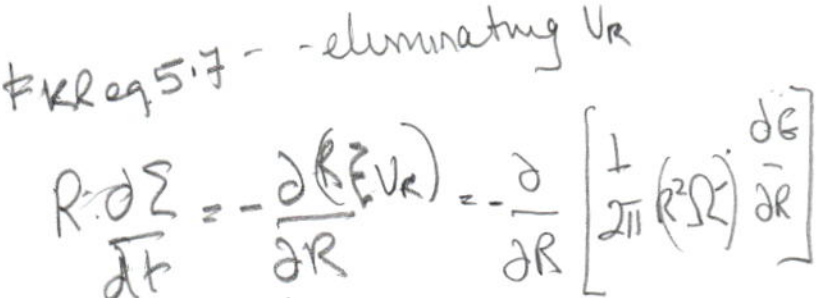

Equation 45 is a non-linear diffusion equation, describing the time evolution of the surface density in the accretion disc due to viscosity. Viscous diffusion is a physical process central to the working of a disc. Before we discuss what 'non-linear diffusion equation' is supposed to mean, and what it tells us about the disc physics, we want you to take a good, hard look at what the innocent word '*gives*' (just before Equation 45) really implies. This will be an exercise in differentiation!

Example 13

Combine Equations 44 and 40 to eliminate the radial drift velocity. Assume that $\partial\Omega/\partial t = 0$. This should give FKR Equation 5.7.

Solution

Equation 44 (FKR Equation 5.4) is

$$R\frac{\partial}{\partial t}(\Sigma R^2 \Omega) + \frac{\partial}{\partial R}(R\Sigma v_R R^2 \Omega) = \frac{1}{2\pi}\frac{\partial G}{\partial R}$$

while Equation 40 (FKR Equation 5.3) is

$$R\frac{\partial \Sigma}{\partial t} + \frac{\partial}{\partial R}(R\Sigma v_R) = 0$$

The first step is to simplify Equation 44. As neither R^2 nor Ω depend on t they can be taken to the front of the derivative. So the first term of Equation 44 can be written as

$$R^2\Omega \times R\frac{\partial}{\partial t}\Sigma$$

But from Equation 40 we know that

$$R\frac{\partial \Sigma}{\partial t} = -\frac{\partial}{\partial R}(R\Sigma v_R) \qquad (46)$$

Using this to replace $R(\partial\Sigma/\partial t)$ we have a modified Equation 44,

$$-R^2\Omega\frac{\partial}{\partial R}(R\Sigma v_R) + \frac{\partial}{\partial R}(R\Sigma v_R R^2\Omega) = \frac{1}{2\pi}\frac{\partial G}{\partial R}$$

With the substitution $a = R^2\Omega$ and $b = R\Sigma v_R$ the left-hand side becomes

$$-a\frac{\partial b}{\partial R} + \frac{\partial}{\partial R}(ab)$$

Using the product rule on the second term gives

$$-a\frac{\partial b}{\partial R} + \left(a\frac{\partial b}{\partial R} + b\frac{\partial a}{\partial R}\right)$$

which simplifies to

$$b\frac{\partial a}{\partial R}$$

Re-substituting for a and b and equating this to the right-hand side of the modified Equation 44 from above gives

$$R\Sigma v_{\mathrm{R}} \frac{\partial}{\partial R}(R^2\Omega) = \frac{1}{2\pi}\frac{\partial G}{\partial R} \qquad (47)$$

This happens to be FKR Equation 5.6. (Note that the prime in FKR Equation 5.6 denotes the operation $\partial/\partial R$.) This can now, finally, be solved for v_{R},

$$v_{\mathrm{R}} = \frac{1}{2\pi}\frac{1}{R\Sigma\dfrac{\partial}{\partial R}(R^2\Omega)}\frac{\partial G}{\partial R} \qquad (48)$$

The last step is to insert this back into Equation 46

$$R\frac{\partial\Sigma}{\partial t} = -\frac{\partial}{\partial R}\left[\frac{1}{2\pi\left(\dfrac{\partial}{\partial R}(R^2\Omega)\right)}\frac{\partial G}{\partial R}\right] \qquad (49)$$

This is FKR Equation 5.7 and so completes the task. ■

None of the steps in the above example are really difficult. It is just a matter of being systematic, carrying out the rules of differentiation and rearranging equations. You do not have to remember all these manipulations – take it as a warm-up exercise for the following questions. Equation 49 leads us to the all-important Equation 45 describing viscous diffusion. Do you like a challenge? Here is one:

Question 31

Substitute the expression for G (FKR Equation 5.5) and the Keplerian value for Ω (Equation 29) into Equation 49 and derive Equation 45.

Hint 1: Calculate the term $\partial G/\partial R$ first; you will have to work out $\partial\Omega/\partial R$ in order to do that.

Hint 2: Then calculate $\partial(R^2\Omega)/\partial R$.

Hint 3: Using these two expressions, simplify the term in square brackets on the right-hand side of Equation 49. Do not carry out the derivative $\partial/\partial R$ that is in front of the square brackets.

Hint 4: Solve the expression you obtain for $\partial\Sigma/\partial t$. ■

Question 32

Using the result of step (3) in Question 31, derive the equation for the radial drift velocity (FKR Equation 5.9) from Equation 48 above. ■

The key result of this subsection is:

> The radial drift of matter in the accretion disc, and hence the rate of change of the local surface density in the disc, can be described as a viscous diffusion process – as encapsulated in Equation 45.

You have already met the phenomenon of *diffusion* in Block 2 (e.g. Section 2.2.2). The random walk of particles constitutes diffusion, and the random walk of photons in an optically thick gas, (i.e. energy transport by radiation) constitutes radiative diffusion. Perhaps the best known example of diffusion is the conduction of heat, such as the loss of heat through the walls of a house. In general terms, diffusion is all about the slow spreading of a physical quantity if this quantity is not evenly distributed. The flow rate of the quantity is proportional to its spatial gradient, i.e. the tendency to spread is particularly large if the distribution is very uneven. In the case of heat conduction you lose heat (thermal energy) through the walls of a house because it is colder outside than inside. The heat flow rate is proportional to the temperature difference. If it is very cold outside you have to turn up the heating system to maintain a comfortable room temperature. The greater the temperature gradient between the room and outside, the greater the heat loss through the wall. (The same is true if there are draughty windows or doors, but in this case the heat loss is due to the exchange of warm air for cold air.)

We now make this a bit more explicit. The rate of change of heat in a small volume is obviously given by the difference between the rate of heat flowing into this volume and the rate of heat flowing out of it. Suppose u denotes the amount of heat energy in a unit volume, (i.e. the heat energy density (measured in erg cm^{-3})), and suppose j_u denotes the heat flow density, i.e. the amount of heat energy flowing through a unit surface per unit time (measured in erg cm^{-2} s^{-1}). Then if the volume is very small, the difference between heat inflow and outflow is just given by the gradient of the flow density. We therefore have for the rate of change of u

$$\frac{\partial u}{\partial t} \propto -\frac{\partial}{\partial x} j_u$$

Here we assumed that the heat flow density is positive if the flow is in the positive x-direction. The minus sign is needed as a decrease of j_u with x ($\partial j_u/\partial x < 0$) leads to an increase of u ($\partial u/\partial t > 0$) at a given point.

The defining characteristic of a diffusion process is that the flow density itself is proportional to the gradient of the heat energy,

$$j_u \propto -\frac{\partial}{\partial x} u$$

■ Why is there a minus sign?

❑ The minus sign is needed because the flow must be in the direction of decreasing heat in order to even out the gradient. ■

Inserting this into the above equation we see that the diffusion process gives rise to an equation which relates the *time* derivative of the heat content to the *second* spatial derivative of the heat content,

$$\frac{\partial u}{\partial t} \propto \frac{\partial^2}{\partial x^2} u \tag{50}$$

An equation of this form is called a **diffusion equation**, whatever the quantity u stands for. The diffusion equation relates the time derivative of a quantity u to its second spatial derivative (see Figure 41). In the example above, u was an energy density (thermal energy per unit volume). Another example is the diffuse mixing of, for example, two different types of gas, where u is the mass density of one of the two gas species.

$$\frac{\partial u}{\partial x} = 0$$

no flow

$$\frac{\partial u}{\partial x} = \text{constant}$$

in = out

$$\frac{\partial u}{\partial x} < 0$$

$$\left.\frac{\partial^2 u}{\partial x^2}\right|_{x_0} > 0$$

in > out

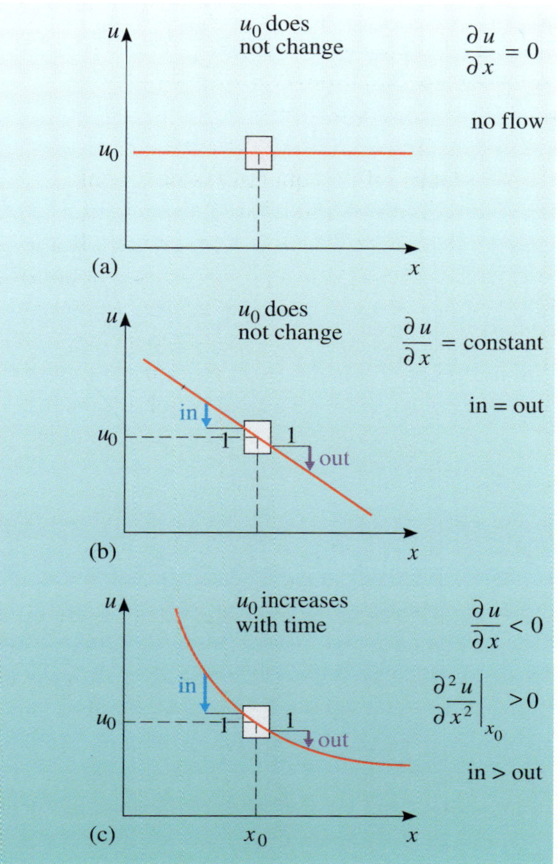

Figure 41 The quantity u changes as a result of diffusion if the second spatial derivative of u, i.e. the curvature of $u(x)$, is non-zero.

Now have a look at our Equation 45 for the surface density Σ in the disc. This is indeed a diffusion equation, as it relates the time derivative of Σ with two spatial derivatives ($\partial/\partial R$ occurs twice) of Σ. Do not worry about the fact that the spatial derivative is more complicated than in the case of u above – this is mainly because of the different geometry. There is one added complication, though, which should worry you. The viscosity ν in Equation 45 can be a function of R, Σ and t itself and therefore can make life rather hard (when spent trying to solve the equation). This is why Equation 45 is, in general, a *non-linear* equation.

But relax, we will start with a particularly simple form of viscosity, a constant viscosity. In this case it is possible to find an analytic solution $\Sigma(R, t)$ of Equation 45. Rather than going through the somewhat tedious algebra needed to find it (even FKR just quotes the result), we instead show you what it looks like in a short animation. But before we can do anything we obviously need to specify how we want the disc to look to begin with – only then does it make sense to ask how it will evolve with time. We choose this **initial condition** to be as simple as possible, but in such a way that it mimics what happens in a real binary system where Roche-lobe overflow has just started. Do you remember the accretion stream animation in Activity 15? There we considered how the gas stream emerging from the inner Lagrangian point eventually settles in a torus at the circularization radius. Accordingly we start here with a very narrow plasma torus at this radius. It will be no surprise to you that this torus will begin to spread and form a more extended disc structure. This is why we also need to specify the fate of material that gets very far out (as well as that which reaches the central star). For simplicity, we assume that the disc is infinite, i.e. so large that the torus can spread forever.

[handwritten margin notes] This is a diffusion equation because it relates time derivative of Σ to two spatial derivatives of Σ.

$$\frac{\partial \Sigma}{\partial t} = \frac{3}{R}\frac{\partial}{\partial R}\left[R^{1/2}\frac{\partial}{\partial R}\left(\nu \Sigma R^{1/2}\right)\right] \quad \text{eq45}$$

is generally a non-linear equation because ν can be a function of (R, Σ, t)

If ν is constant it is possible to find an analytic solution for $\Sigma(R,t)$ (it is too hard to solve.)

Need to look at what the disc looked like in the beginning, & then see how it evolves with time.

Choose initial condition to be as simple as possible but such that it shows what happens in a real binary where Roche flow has just started.

■ Is the disc infinite in a real binary?

❑ No, the disc has to fit inside the Roche lobe of the primary star. In real binaries the disc is truncated by tidal effects from the secondary. The outer disc radius is smaller than the primary's Roche lobe radius, perhaps only half as big. ■

Activity 19 **(20 minutes)**

Viscous diffusion

From *The Energetic Universe* MM guide, start the animation sequence entitled 'Viscous diffusion' showing the viscous spreading of a narrow plasma torus initially located at radius $R_{\rm circ}$, and the direction and magnitude of the drift velocity. (This continues the story started in Activity 15.) Watch the animation sequence and then answer Question 33.

Keywords: none ■

Question 33

(a) Describe the behaviour of the viscously spreading ring of gas. (b) What is the significance of the point where the radial drift velocity $v_{\rm R}$ changes sign? (c) What happens to the drift velocity $v_{\rm R}$ at a fixed radius $R > R_{\rm circ}$? (d) After a very long time, how are the mass and angular momentum distributed in the disc? ■

4.1.4 The viscous time

The narrow plasma torus you have investigated in the last activity spreads on a characteristic timescale. This so-called **viscous time**, or **radial drift timescale**, turns out to be $t_{\rm visc} \sim R^2/\nu$. It is possible to see this directly from the functional form of the analytic solution whose behaviour we have shown you in the animation. But an estimate for the timescale can also be found in a rather simple way from the viscous diffusion equation itself, without actually solving it rigorously. The way this is done is of general use and usually a great help when you are faced with a rather involved physical situation with a correspondingly complicated equation at hand. So let us sit back for a few minutes and consider this trick in some detail. It helps if we rearrange the viscous diffusion equation slightly. Take it as another example of applying the rules of differentiation.

Example 14

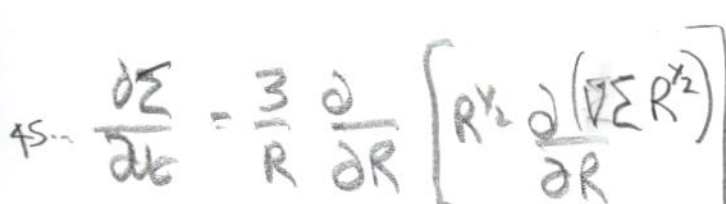

Rewrite the right-hand side of Equation 45 as a sum of terms involving the first and second spatial derivatives of Σ. Assume a constant viscosity.

Solution

As ν = constant, we can move it to the front, and Equation 45 becomes

$$\frac{\partial \Sigma}{\partial t} = \frac{3\nu}{R}\frac{\partial}{\partial R}\left\{R^{1/2}\frac{\partial}{\partial R}(\Sigma R^{1/2})\right\}$$

Using the product rule on the inner derivative gives

$$\begin{aligned}\frac{\partial \Sigma}{\partial t} &= \frac{3\nu}{R}\frac{\partial}{\partial R}\left\{R^{1/2}\left[\Sigma\frac{\partial}{\partial R}(R^{1/2}) + \frac{\partial \Sigma}{\partial R}R^{1/2}\right]\right\}\\ &= \frac{3\nu}{R}\frac{\partial}{\partial R}\left\{R^{1/2}\left[\Sigma\frac{1}{2R^{1/2}} + \frac{\partial \Sigma}{\partial R}R^{1/2}\right]\right\}\end{aligned}$$

Factoring in $R^{1/2}$ gives

$$\frac{\partial \Sigma}{\partial t} = \frac{3\nu}{R}\frac{\partial}{\partial R}\left\{\frac{\Sigma}{2} + \frac{\partial \Sigma}{\partial R}R\right\}$$

Using the sum rule this becomes

$$\frac{\partial \Sigma}{\partial t} = \frac{3\nu}{R}\left\{\frac{\partial}{\partial R}\left(\frac{\Sigma}{2}\right) + \frac{\partial}{\partial R}\left(\frac{\partial \Sigma}{\partial R}R\right)\right\}$$

Now using the product rule again gives

$$\frac{\partial \Sigma}{\partial t} = \frac{3\nu}{R}\left\{\frac{1}{2}\frac{\partial \Sigma}{\partial R} + \frac{\partial^2 \Sigma}{\partial R^2}R + \frac{\partial \Sigma}{\partial R}\right\} = \frac{3\nu}{R}\left\{\frac{3}{2}\frac{\partial \Sigma}{\partial R} + \frac{\partial^2 \Sigma}{\partial R^2}R\right\}$$

So we have finally

$$\frac{\partial \Sigma}{\partial t} = \frac{9\nu}{2R}\frac{\partial \Sigma}{\partial R} + 3\nu\frac{\partial^2 \Sigma}{\partial R^2} \tag{51}$$

Of course, this is just an alternative way to write Equation 45 in the case of constant viscosity. ■

Now a quick and dirty way to estimate the viscous time would be to just 'ignore' the ∂ symbols. Then Equation 51 suggests a relation like

$$\frac{\Sigma}{t} \sim \nu\frac{\Sigma}{R^2} \quad \text{or} \quad t \sim \frac{R^2}{\nu}$$

Note that we have dropped numerical constants of order unity, i.e. constants with a value not too different from 1, between about 0.3 and 5, say. What we have just done can also be expressed in a more sophisticated way. On *dimensional grounds* we expect $t_{\text{visc}} \sim R^2/\nu$, as this is the simplest possible combination of quantities in the viscous diffusion equation with the dimension of time.

■ Using the α-viscosity, verify that R^2/ν has the dimension of time.

❑ The α-viscosity is $\nu = \alpha c_s H$. As α is dimensionless, ν has the unit $(\text{m s}^{-1}) \times \text{m} = \text{m}^2\,\text{s}^{-1}$. So R^2/ν has the unit $\text{m}^2/(\text{m}^2\,\text{s}^{-1}) = \text{s}$. ■

Given the simplicity of this 'method' it is surprising that it works at all. But beware – a too simple-minded dimensional analysis could go badly wrong as well! It is a good idea to check the *guess* obtained in this way against a slightly more rigorous – though still approximate – approach. We do this now.

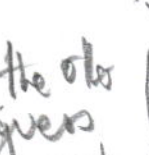

We replace the derivatives in Equation 51 with *typical values*, or *order-of-magnitude* estimates. In a sense, we will be using the very definition of a derivative (as given, for example, in Equation 74 of Block 1) backwards. First we replace $\partial\Sigma/\partial t$ at radius R with the value of Σ at this radius, divided by a representative time over which Σ changes, the viscous time t_{visc}

$$\frac{\partial \Sigma}{\partial t} \approx \frac{\Sigma}{t_{\text{visc}}}$$

Next we replace the spatial gradient $\partial\Sigma/\partial R$ with Σ, divided by a length scale l that characterizes the spatial gradient (see Figure 42 overleaf). Then Equation 51 becomes

$$\frac{\Sigma}{t_{\text{visc}}} \approx \frac{9\nu}{2R}\frac{\Sigma}{l} + 3\nu\frac{\Sigma}{l^2}$$

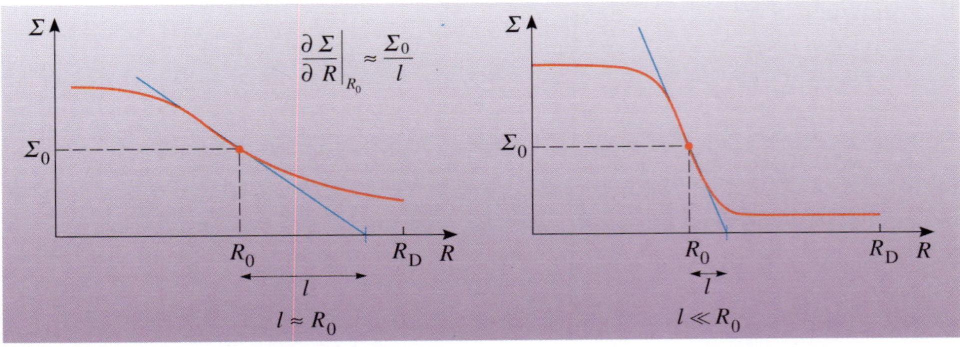

Figure 42 Order-of-magnitude estimates for gradients in differential equations. Shown are two different surface density profiles in accretion discs. The slope $\partial\Sigma/\partial R$ of the curve at radius R_0 is approximated as Σ_0/l. R_D is the outer disc radius.

Dropping numerical factors of order unity and rearranging we have approximately

$$\frac{\Sigma}{t_{\text{visc}}} \approx \nu\Sigma\left(\frac{1}{Rl}+\frac{1}{l^2}\right) = \frac{\nu\Sigma}{l^2}\left(1+\frac{l}{R}\right) \qquad (52)$$

If $l \ll R$ then the term in brackets is close to unity, and the expression Equation 52 can be rearranged to give $t_{\text{visc}} \approx l^2/\nu$. If, on the other hand, we have $l \approx R$ we find from Equation 52

$$\frac{\Sigma}{t_{\text{visc}}} \approx \frac{\nu\Sigma}{R^2}\left(1+\frac{R}{R}\right)$$

and hence the old result $t_{\text{visc}} \sim R^2/\nu$ (again dropping the numerical factor 2). This shows that density enhancements involving sharp gradients, with $l \ll R$, diffuse more quickly than smoother density enhancements. Clearly, the 'quick and dirty' method from above failed to pick this up.

Shall we try again with another example for an order-of-magnitude estimate?

Example 15

Find an order-of-magnitude estimate for the radial drift velocity (FKR Equation 5.9). Assume a constant viscosity.

Solution

FKR Equation 5.9 reads

$$\upsilon_R = -\frac{3}{\Sigma R^{1/2}}\frac{\partial}{\partial R}(\nu\Sigma R^{1/2})$$

The viscosity can be moved to the front of the expression. We use the product rule to find

$$\upsilon_R = -\frac{3\nu}{\Sigma R^{1/2}}\left(R^{1/2}\frac{\partial\Sigma}{\partial R} + \Sigma\frac{\partial(R^{1/2})}{\partial R}\right)$$

or

$$\upsilon_R = -\frac{3\nu}{\Sigma R^{1/2}}\left(\frac{\partial\Sigma}{\partial R}R^{1/2} + \Sigma\frac{1}{2R^{1/2}}\right)$$

Rearranging this we obtain

$$\upsilon_R = -\frac{3\nu}{\Sigma}\frac{\partial\Sigma}{\partial R} - \frac{3\nu}{2R}$$

If l is the typical length scale associated with the gradient of Σ, and $l \approx R$, then both terms in the sum on the right-hand side have the same order of magnitude, ν/R. Hence we expect $\upsilon_R \sim -\nu/R$, or, if we are just interested in the magnitude (not the sign), $\upsilon_R \sim \nu/R$. ■

Let us recapitulate. We found $t_{visc} \sim R^2/\nu$ and $\upsilon_R \sim \nu/R$. If we combine these to eliminate the viscosity we also have $t_{visc} \sim R/\upsilon_R$! You should feel very pleased with yourself. This is exactly what you would write down if somebody asked you to define the radial drift timescale: drift velocity = distance travelled ÷ the time it took, so the drift time is simply this distance over velocity.

From the amount of time we allowed you to spend on the viscous time you might have already guessed that this is something rather important. So much so that we repeat it yet again

$$t_{visc} \sim R^2/\nu \quad \text{viscous time} \tag{53}$$

$$\upsilon_R \sim \nu/R \quad \text{radial drift velocity} \tag{54}$$

4.2 Steady-state discs

After all this effort invested in deriving the equations that govern the radial structure of thin discs, what can we actually learn from them? Do you remember the approach we planned to take? One rather successful example of insights gained by simplifications has been the evolution of the plasma torus with constant viscosity. We now turn to another simplification with a much greater impact: we consider *steady-state* accretion, i.e. *time-independent* accretion. Do not confuse this with the gas being *at rest*. All we are saying is that the disc appears the same whenever we look at it. The plasma in the disc will still swirl around the central object on Keplerian orbits. For the model, we require that none of the quantities describing the steady-state disc depend explicitly on time, so all partial derivatives of the form $\partial/\partial t$ will vanish. This greatly simplifies the disc equations! In fact, they will no longer be partial differential equations at all, they reduce to ordinary differential equations, with R as the only independent variable. We shall learn a great deal about discs by just considering steady-state accretion. In fact, we shall even understand time-dependent discs (in Section 8) largely by insights we gain from these steady-state discs.

Activity 20 **(15 minutes)**

Integrating the disc equations

You should now read Section 5.3 of FKR up to and including FKR Equation 5.19.

As before, you will find assistance below. First read FKR, but do not dwell on any of the equations or manipulations of equations which you do not understand. Then read the hints below, and – where appropriate – return to FKR and read it again.

Note that in this reading the symbol R_* denotes the radius of the compact star, *not* the radius of the donor star as in Section 2. This makes sense as we are now concerned with the accretion disc which is centred on the accretor. In fact, in most of our analysis of the physics of accretion discs the presence of a companion star has been neglected.

Keywords: **steady-state disc, boundary layer** ■

The assumption of steady-state accretion allows one to *integrate* the accretion disc equations, i.e. to find solutions of these equations. In the case of the conservation of mass we know the integral already. It is the mass accretion rate $\dot{M}$ which FKR introduces in FKR Equation 5.14. We have defined this quantity already in Equation 42 above, as the *local* mass accretion rate through the disc. In steady-state discs the radial mass flow rate has to be the same everywhere in the disc, at all times – otherwise mass would pile up or deplete in certain disc rings, in clear conflict with the assumption of a steady state. This constant mass accretion rate must also equal the rate at which the Roche-lobe filling donor star feeds mass into the disc. Hence the mass transfer rate is the same as the mass accretion rate.

The second integral is the one of the angular momentum equation (Equation 44 and FKR 5.4)

$$R\frac{\partial}{\partial t}(\Sigma R^2 \Omega) + \frac{\partial}{\partial R}(R\Sigma v_{\rm R} R^2 \Omega) = \frac{1}{2\pi}\frac{\partial G}{\partial R}$$

As this is the first time in this block that you encounter an explicit integration of a differential equation we will spend a few minutes explaining the procedure. For steady-state discs the term with $\partial/\partial t$ vanishes, so we are left with

$$\frac{\partial}{\partial R}(R\Sigma v_{\rm R} R^2 \Omega) = \frac{1}{2\pi}\frac{\partial G}{\partial R} \tag{55}$$

Luckily, both sides of the equation are derivatives with respect to R. Note that since we are dealing only with derivatives with respect to R it does not matter if we write $\mathrm{d}/\mathrm{d}R$ or $\partial/\partial R$; they are synonymous. To 'get rid of' the derivatives, we apply the inverse operation, an integration over R, on both sides of the equation. Quite generally, for any function $f(R)$, we have the identity

$$\int \frac{\partial f(R)}{\partial R}\,\mathrm{d}R = f(R) + C \tag{56}$$

with C being an arbitrary constant. The ambiguity expressed by the integration constant C appears as we did not specify the integration boundaries: for any C we have

$$\frac{\partial}{\partial R}(f(R) + C) = \frac{\partial f(R)}{\partial R}$$

We can now integrate Equation 55 by applying the rule expressed in Equation 56 on both the left- and right-hand side of Equation 55. This gives

$$R\Sigma v_{\rm R} R^2 \Omega = \frac{G}{2\pi} + \frac{C}{2\pi} \tag{57}$$

where we have combined the two integration constants into one new constant $C/(2\pi)$. This is purely for convenience – as long as there is a constant in the equation it is not important what form it has.

Question 34

Show that Equation 57 reduces to FKR Equation 5.15 if the standard form for G (Equation 36) is used. ■

Clearly, to make use of the integrated equation we need to know what the integration constant C is. As Equation 57 is valid everywhere in the disc we can determine the value of C at any disc annulus we like. A clever choice is a point where $\mathrm{d}\Omega/\mathrm{d}R = 0$, because this implies that one term in FKR Equation 5.15 conveniently vanishes.

But where is $\mathrm{d}\Omega/\mathrm{d}R = 0$? To answer this we have to make a short detour and introduce the boundary layer.

4.2.1 The boundary layer

The boundary layer is the innermost region of the accretion flow, the transition zone between the Keplerian disc and the surface of the accreting star. Usually this star rotates with an angular speed Ω_* well below the local Keplerian value at the surface of the star,

$$\Omega_{\mathrm{K}}(R_*) = \sqrt{\frac{GM}{R_*^3}} \tag{58}$$

Conversely, we know that at large radii R, far away from the star, the angular speed of the material in the disc is indeed just the Kepler rate $\Omega_{\mathrm{K}}(R) = (GM/R^3)^{1/2}$. Clearly we can expect that the angular speed Ω of the disc plasma does not suddenly (discontinuously) drop from Ω_{K} to the smaller Ω_*. Rather, we expect a smooth transition as shown in Figure 43. Hence Ω must start to deviate from the Kepler rate once the radius is smaller than some value $R = R_* + b$. The region of the accretion flow between the surface of the accretor and the radius $R_* + b$ is the boundary layer. In other words, b is the radial width of the boundary layer.

On Keplerian orbits the centrifugal force just balances the gravitational force that pulls the disc material inwards towards the central accretor. This balance breaks down in the boundary layer: the centrifugal force is much smaller than the gravitational force. Instead, gravity is roughly balanced by a radial pressure gradient.

It turns out that this requires that the boundary layer is not very extended in radial direction – this allows for a larger pressure gradient. More specifically, the width b is small compared to both the radius of the accreting star and the height of the disc just outside the boundary layer. FKR shows this in detail in FKR Section 6.2 – due to time constraints, this is not a compulsory reading for this course. So you have to trust us on this!

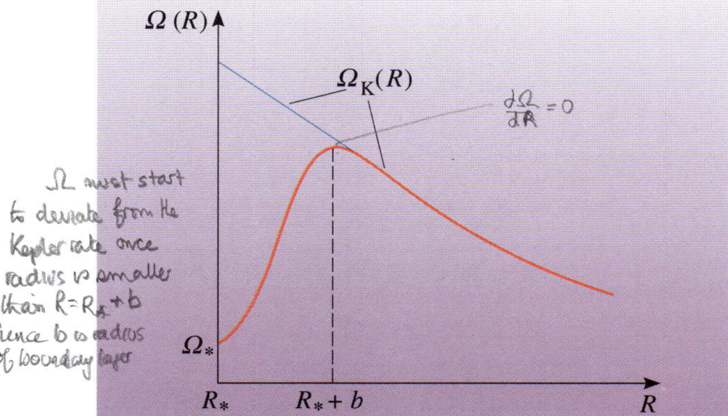

Figure 43 The angular speed near the inner edge of an accretion disc around a star with surface angular speed Ω_* less than the value given in Equation 58.

4.2.2 The steady-state surface density

We now resume the discussion of how to determine the integration constant C in Equation 57. We have just identified the outer edge of the boundary layer as the point where $\mathrm{d}\Omega/\mathrm{d}R \approx 0$. The angular speed itself is still approximately Keplerian at this point, and, up to terms of order $O(b/R_*)$ (see the box on 'Order'), given by the rate defined in Equation 58.

ORDER!

The term $O(x)$ reads *terms of order x or higher* and denotes a sum of the form

$$O(x) = c_1 x + c_2 x^2 + c_3 x^3 + \ldots$$

with constant coefficients c_1, c_2, c_3, ... The sum is a series of terms with increasing power in x, but the *lowest* power of x is 1. The rationale behind this is that if $x \ll 1$, then each of the terms is small, in fact negligibly small against a term of order unity. More generally, $O(x^n)$ denotes terms of the form $c_n x^n + c_{n+1} x^{n+1} + c_{n+2} x^{n+2} + \ldots$ (n is an integer). The use of this notation usually implies that anything subsumed in $O(x^n)$ will be neglected.

Question 35

Show that the angular velocity at $R = R_* + b$ is given by the rate defined in Equation 58, up to terms of order $O(b/R_*)$. *Hints*: (a) Write down the Keplerian angular speed at radius $R = R_* + b$ and factor out the term $(GM/R_*^3)^{1/2}$. (b) Expand the second term in a Taylor series in terms of b/R_* (see, for example, Example 20 in Section 3.7 of Block 1). (c) Compare your result with FKR Equation 5.17. ■

As shown by FKR, the integration constant then becomes

$$C = -\dot{M}(GMR_*)^{1/2} \tag{59}$$

Using this value for C we now wish to derive FKR Equation 5.19 from FKR Equation 5.15,

$$-\nu\Sigma\Omega' = \Sigma(-\upsilon_R)\Omega + \frac{C}{2\pi R^3}$$

We use Equation 59 for C, the Keplerian value for the angular speed Ω (Equation 29),

$$\Omega = \left(\frac{GM}{R^3}\right)^{1/2}$$

and the corresponding derivative of Ω with respect to R,

$$\Omega' = -\frac{3}{2}\left(\frac{GM}{R^5}\right)^{1/2} = -\frac{3}{2R}\left(\frac{GM}{R^3}\right)^{1/2}$$

Inserting all this into FKR Equation 5.15 gives

$$-\nu\Sigma\frac{-3}{2R}\left(\frac{GM}{R^3}\right)^{1/2} = \Sigma(-\upsilon_R)\left(\frac{GM}{R^3}\right)^{1/2} - \frac{\dot{M}(GMR_*)^{1/2}}{2\pi R^3}$$

Solving for $\nu\Sigma$ we obtain

$$\nu\Sigma = \frac{2R}{3}\left(\frac{GM}{R^3}\right)^{-1/2} \times \left[\Sigma(-\upsilon_R)\left(\frac{GM}{R^3}\right)^{1/2} - \frac{\dot{M}R_*^{1/2}}{2\pi R^{3/2}}\left(\frac{GM}{R^3}\right)^{1/2}\right]$$

which simplifies to

$$\nu\Sigma = \frac{2}{3}R\Sigma(-\upsilon_R) - \frac{2}{3}\frac{\dot{M}}{2\pi}\left(\frac{R_*}{R}\right)^{1/2} \tag{60}$$

Using Equation 42 we find that

$$R(-\upsilon_R)\Sigma = \frac{\dot{M}}{2\pi}$$

so that Equation 60 becomes

$$\nu\Sigma = \frac{2}{3}\frac{\dot{M}}{2\pi} - \frac{2}{3}\frac{\dot{M}}{2\pi}\left(\frac{R_*}{R}\right)^{1/2}$$

i.e.

$$\nu\Sigma = \frac{\dot{M}}{3\pi} - \frac{\dot{M}}{3\pi}\left(\frac{R_*}{R}\right)^{1/2}$$

or finally

$$\nu\Sigma = \frac{\dot{M}}{3\pi}\left[1 - \left(\frac{R_*}{R}\right)^{1/2}\right] \qquad \text{(FKR 5.19) (61)}$$

■ Describe the radial dependence in Equation 61 far away from the central star, and close to the surface of it.

□ Far away from the accreting star, i.e. for $R \gg R_*$, the term $(R_*/R)^{1/2}$ is small compared with 1 and can be neglected. This means that the product $\nu\Sigma$ of viscosity and surface density is constant throughout the outer disc. This is no longer true close to the surface of the accreting star. When R approaches R_* the product $\nu\Sigma$ approaches 0. ■

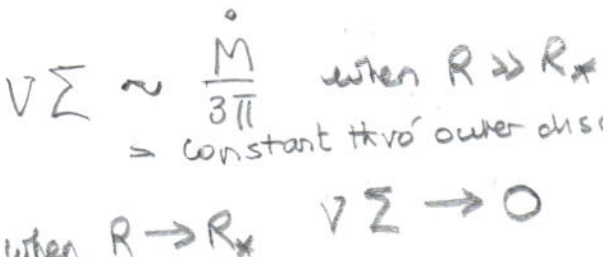

Activity 21 **(30 minutes)**

$\nu\Sigma$ in a steady-state disc

Use a spreadsheet to verify the above statements on the radial dependence of $\nu\Sigma$. Plot the function $1 - (R_*/R)^{1/2}$ in the radial range from $R = R_*$ to $R = 100 \times R_*$.

Keywords: none ■

Equation 61 shows that over a large region of a steady-state accretion disc $\nu\Sigma$ = constant and so the surface density is inversely proportional to the viscosity. Whatever the physical mechanism providing the viscosity is, in a steady-state disc with a given mass accretion rate the disc will adjust its local structure (surface density, temperature) such that Equation 61 is fulfilled.

Question 36

Interpret the physical significance of the result $\nu\Sigma \propto \dot{M}$ by using order-of-magnitude estimates for the viscous time and radial drift velocity. *Hint*: Use Equation 53 to eliminate the viscosity in the product $\nu\Sigma$. (a) First express $\nu\Sigma$ in terms of the viscous time. How does the resulting expression relate to $\dot{M}$? (Compare dimensions.) (b) Then express $\nu\Sigma$ in terms of the radial drift velocity. Considering Equation 42, how does the resulting expression relate to $\dot{M}$? ■

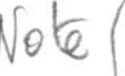

You worked so hard to obtain Equation 61 that it is time to show you what amazing things it tells us about steady-state accretion discs. As a first application, we now calculate how the disc luminosity varies with distance from the central star.

4.2.3 The accretion disc luminosity

Accretion disc luminosity

Read the two paragraphs in FKR that contain Equations 5.20–5.22. This begins at the bottom of page 85 with 'The expression (5.19) …' and ends with 'We shall discuss the boundary layer in the next chapter' on page 86. In fact, you know some of this already from above, and you will hear a bit more about the boundary layer at the end of this subsection.

As before, you will find assistance below.

Keywords: **disc luminosity** ■

The key ingredient for calculating the luminosity of a disc ring between radii R_1 and R_2 is the viscous dissipation rate. It is assumed that the disc radiates the energy at the same rate that viscous dissipation generates it. Although the dissipation rate generally depends on the viscosity, it was possible to eliminate the viscosity using the steady-state relation Equation 61.

Question 37

How can FKR Equation 5.20 be derived from FKR Equation 4.30? ■

Example 16

Show all steps of the integration that lead to the expression for the luminosity of a disc ring between radii R_1 and R_2 (FKR Equation 5.21).

Solution

The luminosity (energy generated per unit time) is given by the integral of the dissipation rate per unit surface area,

$$D(R) = \frac{3GM\dot{M}}{8\pi R^3}\left[1-\left(\frac{R_*}{R}\right)^{1/2}\right]$$

(FKR Equation 5.20), over the surface of the disc ring,

$$L(R_1, R_2) = 2\times \int_{R_1}^{R_2} D(R)2\pi R\,\mathrm{d}R$$

Here the azimuthal part of the surface integral has already been carried out, giving the factor 2π in the integrand (see Block 1, Section 3.15, Example 28). The factor 2 in front of the integral accounts for the fact that the disc has two faces. Inserting D and collecting all constants in front of the integral gives

$$L(R_1, R_2) = \frac{3GM\dot{M}}{2}\int_{R_1}^{R_2}\left[1-\left(\frac{R_*}{R}\right)^{1/2}\right]\left(\frac{1}{R^2}\,\mathrm{d}R\right)$$

We make the substitution $y = R_*/R$. As

$$\frac{\mathrm{d}y}{\mathrm{d}R} = -\frac{R_*}{R^2}$$

Handwritten margin notes:

Activity 22
FKR p 85, 86

accretion disc L — the E an accretion disc radiates per unit time. For steady state, geometrically thin, optically thick, infinite disc with a non-rotating central accelerator with mass M radius R_* and accretion rate \dot{M} the accretion disc L is

$$L_{disc} = \frac{GM\dot{M}}{2R_*}$$

(37) $D(R) = \frac{9}{8}(\nu\Sigma)\frac{GM}{R^3}\cdots$ (4.30)

$D(R) = \frac{3GM\dot{M}}{8\pi R^3}\left(1-\left(\frac{R_*}{R}\right)^{1/2}\right)$ (5.20)

the steady state relation eq. 61

$\nu\Sigma = \frac{\dot{M}}{3\pi}\left[1-\left(\frac{R_*}{R}\right)^{1/2}\right]$ Insert this in 4.30

$D(R) = \frac{9}{8}\frac{\dot{M}}{3\pi}\left[1-\left(\frac{R_*}{R}\right)^{1/2}\right]\frac{GM}{R^3}$

$= \frac{3GM\dot{M}}{8\pi R^3}\left[1-\left(\frac{R_*}{R}\right)^{1/2}\right]$

This is eq. 5.20.

$\nu\Sigma = \frac{\dot{M}}{3\pi}\left[1-\sqrt{R_*/R}\right]$ (61)

$\left(-\frac{\mathrm{d}y}{R_*}\right) = \left(\frac{\mathrm{d}R}{R^2}\right)$

we can replace (dR/R^2) with $(-dy/R_*)$ and obtain

$$L(R_1, R_2) = -\frac{3GM\dot{M}}{2R_*}\int_{y_1}^{y_2}[1 - y^{1/2}]\,dy \qquad (62)$$

where $y_1 = R_*/R_1$ and $y_2 = R_*/R_2$. The integral can be carried out as follows:

$$\int_{y_1}^{y_2}[1 - y^{1/2}]\,dy = \int_{y_1}^{y_2}dy - \int_{y_1}^{y_2}y^{1/2}\,dy$$

$$= y_2 - y_1 - \left[\frac{2}{3}y^{3/2}\right]_{y_1}^{y_2} = y_2 - y_1 - \frac{2}{3}y_2^{3/2} + \frac{2}{3}y_1^{3/2}$$

$$= -\left[y_1\left(1 - \frac{2}{3}y_1^{1/2}\right) - y_2\left(1 - \frac{2}{3}y_2^{1/2}\right)\right]$$

Inserting this for the integral in Equation 62 above, re-substituting R_*/R_1 for y_1, and R_*/R_2 for y_2, gives

$$L(R_1, R_2) = -\frac{3GM\dot{M}}{2R_*}\times(-1)\times\left\{\frac{R_*}{R_1}\left[1 - \frac{2}{3}\left(\frac{R_*}{R_1}\right)^{1/2}\right] - \frac{R_*}{R_2}\left[1 - \frac{2}{3}\left(\frac{R_*}{R_2}\right)^{1/2}\right]\right\}$$

Cancelling the leading R_* finally gives FKR Equation 5.21,

$$L(R_1, R_2) = \frac{3GM\dot{M}}{2}\left\{\frac{1}{R_1}\left[1 - \frac{2}{3}\left(\frac{R_*}{R_1}\right)^{1/2}\right] - \frac{1}{R_2}\left[1 - \frac{2}{3}\left(\frac{R_*}{R_2}\right)^{1/2}\right]\right\} \qquad (63)$$ ■

Question 38

Use Equation 63 to calculate the luminosity of the whole disc.

Hint: Set $R_1 = R_*$ and $R_2 \to \infty$. ■

By answering the last question you should have found that the integral luminosity of a geometrically thin, optically thick steady-state accretion disc is

$$L_{\text{disc}} = \frac{1}{2}\frac{GM\dot{M}}{R_*}$$

This is just $0.5 \times L_{\text{acc}}$ (see Equation 2). The other half of the accretion luminosity is in fact radiated by the boundary layer!

It is worth spending some time thinking about how this comes about. The total energy of a plasma blob with mass m on a Kepler orbit with radius R around the accreting star with mass M is the sum of the gravitational potential energy and kinetic energy,

$$E_{\text{TOT}} = E_{\text{GR}} + E_{\text{KE}} = -\frac{GMm}{R} + \frac{1}{2}mv^2$$

The orbital speed is $v = (GM/R)^{1/2}$ (Equation 1), so the kinetic energy is

$$E_{\text{KE}} = \frac{1}{2}m\left(\sqrt{\frac{GM}{R}}\right)^2 = \frac{1}{2}\frac{GMm}{R}$$

Therefore, we have

$$E_{\text{KE}} = -\tfrac{1}{2}E_{\text{GR}} \qquad (64)$$

Incidentally, this is just another form of the virial theorem which you have met in Section 1.2.4 of Block 2. As a consequence of Equation 64, we also have $E_{\text{TOT}} = \frac{1}{2} E_{\text{GR}}$.

■ Why?

☐ Because $E_{\text{TOT}} = E_{\text{GR}} + E_{\text{KE}} = E_{\text{GR}} + (-\frac{1}{2}) \times E_{\text{GR}} = \frac{1}{2} E_{\text{GR}}$ ■

When the plasma blob arrives at the inner edge of the disc ($R = R_* + b \approx R_*$) it has lost the energy ΔE_{TOT}

$$\Delta E_{\text{TOT}} = \frac{1}{2} E_{\text{GR}}(R \to \infty) - \frac{1}{2} E_{\text{GR}}(R_*) = 0 - \frac{1}{2}\frac{(-GMm)}{R_*} = \frac{GMm}{2R_*}$$

At this point the mass blob still has the kinetic energy $E_{\text{KE}} = GMm/(2R_*)$. In the boundary layer, the plasma blob slows down and eventually comes to rest on the surface of the accreting star. Hence in the boundary layer the plasma blob loses all its kinetic energy

$$\Delta E_{\text{KE}} = GMm/2R_*$$

The disc luminosity is proportional to ΔE_{TOT} while the boundary layer luminosity is proportional to ΔE_{KE}, hence $L_{\text{disc}}/L_{\text{BL}} = \Delta E_{\text{TOT}}/\Delta E_{\text{KE}} = 1$ and therefore $L_{\text{disc}} = L_{\text{BL}}$. The total accretion luminosity is $L_{\text{acc}} = L_{\text{disc}} + L_{\text{BL}} = 2 \times L_{\text{disc}}$.

As the boundary layer is so much smaller than the disc but nonetheless has the same luminosity, it inevitably must be hotter than the disc.

■ Why?

☐ Because the radiant flux (= luminosity per unit area) is proportional to the temperature to the fourth power. This is the Stefan–Boltzmann law. ■

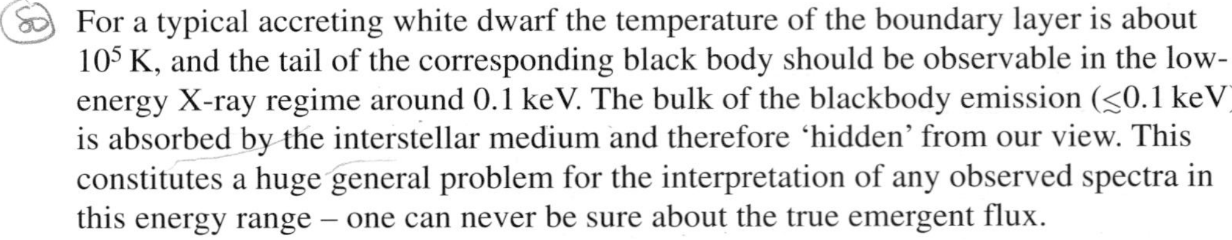

For a typical accreting white dwarf the temperature of the boundary layer is about 10^5 K, and the tail of the corresponding black body should be observable in the low-energy X-ray regime around 0.1 keV. The bulk of the blackbody emission ($\lesssim$0.1 keV) is absorbed by the interstellar medium and therefore 'hidden' from our view. This constitutes a huge general problem for the interpretation of any observed spectra in this energy range – one can never be sure about the true emergent flux.

The name game: part 4

Skim-read Section 4 and make a list of all symbols that denote *new* variables and *new* quantities, and note down what they mean. Flag up those symbols that have (unfortunately) multiple uses.

Once you have compiled your list and explained the meaning of all entries, compare it to the one given in Appendix A5 at the end of the Study Guide.

4.3 Summary of Section 4

1 A good physical model should capture the essential physical ingredients of the process or system, but not dwell on unnecessary details.

2 The radial structure of geometrically thin accretion discs is determined by the conservation of mass

$$R\frac{\partial \Sigma}{\partial t} + \frac{\partial}{\partial R}(R\Sigma \upsilon_{\text{R}}) = 0$$

and the conservation of angular momentum

$$R\frac{\partial}{\partial t}(\Sigma R^2 \Omega) + \frac{\partial}{\partial R}(R\Sigma v_R R^2 \Omega) = \frac{1}{2\pi}\frac{\partial G}{\partial R}$$

where Σ is the surface density, Ω the angular speed, G the viscous torque and v_R the radial drift velocity.

3 The local mass accretion rate

$$\dot{M}(R, t) = -2\pi R v_R \Sigma$$

in the disc is the amount of mass that flows per unit time through the boundary at radius R between adjacent disc annuli.

4 Viscosity causes viscous diffusion of the surface density. The equation describing viscous diffusion is

$$\frac{\partial \Sigma}{\partial t} = \frac{3}{R}\frac{\partial}{\partial R}\left[R^{1/2}\frac{\partial}{\partial R}(\nu \Sigma R^{1/2})\right]$$

5 An equation of the form

$$\frac{\partial u}{\partial t} \propto \frac{\partial^2}{\partial x^2}u$$

is called diffusion equation. It describes flow effects that even out spatial gradients. The flow strength scales with the steepness of the spatial gradient.

6 Density enhancements involving sharp gradients diffuse more quickly than smoother density enhancements. The viscous time is $t_{\text{visc}} \sim R^2/\nu$, the radial drift velocity is $v_R \sim \nu/R$, where ν is the viscosity.

7 Order-of-magnitude estimates allow one to obtain characteristic timescales and dimensions for physical systems and processes from otherwise complicated looking differential equations describing these systems and processes.

8 The assumption of steady-state accretion simplifies the integration of the radial disc structure equations. The integral of the equation of mass conservation is the mass accretion rate $\dot{M}$. In a steady-state disc the mass accretion rate is constant and equal to the mass transfer rate from the secondary star.

9 A full integration of the equation describing the conservation of angular momentum requires knowledge of the structure of the boundary layer, the transition zone between the Keplerian accretion disc and the central accreting star.

10 The radial extent of the boundary layer is very small. Pressure forces balance gravity in radial direction.

11 A steady-state Keplerian disc around a slowly rotating star will adjust its local structure such that

$$\nu\Sigma = \frac{\dot{M}}{3\pi}\left[1 - \left(\frac{R_*}{R}\right)^{1/2}\right]$$

whatever the physical mechanism of the viscosity. Here R_* is the radius of the accreting object.

12 If the accretor is a slow rotator the luminosity of the whole disc is equal to the luminosity of the boundary layer, i.e. each is just half of the accretion luminosity.

5 MORE ON 'DISCS AT PEACE' ...

We continue our journey through the world of steady-state accretion discs in this section. As you will see shortly, more and more of the sometimes rather abstract concepts we have developed in the last section will now come to life. We begin by considering how the surface temperature in the disc varies with distance – allowing us to *predict* what the spectrum of such an accretion disc should look like!

5.1 Colourful discs: the spectral energy distribution

For the first time in this block we are now in a position to check our theoretical considerations for steady-state discs by experiment. But how? Well, astronomers gain experimental data on celestial objects by collecting the electromagnetic radiation these objects emit. In particular, they look at the spectral distribution of this radiation. So, given the disc properties we have deduced so far, would you recognize the spectral signature of an accretion disc if you pointed your telescope towards one?

To make sure that you would, we had better calculate the emitted spectrum first. As usual we do this in an approximate way. The concept is simple enough. We know that *in equilibrium* the flux emerging from a disc annulus must equal the rate at which viscous dissipation deposits energy into this disc ring. We have just used this idea to calculate the luminosity of the disc ring. If the disc is also **optically thick** (i.e. opaque), we expect that the flux emerging from the ring is that of a black body, characterized by the surface temperature – or, to be precise, the effective temperature – of the disc ring. (See also the discussion in Block 1, Section 2.6.) As you will see this temperature rises towards the centre of the disc, so that the resulting spectrum is a superposition of many Planck functions with different temperatures.

Activity 23 **(20 minutes)**

The accretion disc continuum spectrum

You should now read Section 5.5 of FKR up to and including the first sentence on page 92 ('The spectrum given by (5.45) is shown in Fig. 5.2.').

A word of warning: so far, FKR has used the symbol v to denote the kinematic viscosity. In this section of FKR, v stands for the frequency of an electromagnetic wave (as it did in Block 1). The only justification for this is that nowhere in FKR do viscosity and frequency occur side by side, and that it is indeed common practice to use v as a symbol for both viscosity and frequency. Similarly, the distance to the accretion disc from Earth will appear as D, although this symbol has been used before to denote the dissipation rate.

Keywords: **black body, accretion disc spectrum** ■

The continuum spectrum of an optically thick steady-state disc is expressed in terms of the flux F_v a detector would pick up on Earth. The flux F_v is actually the amount of energy from the disc that a detector registers per unit time and unit surface area in a unit frequency interval (Block 1, Section 2.6). This must vary as $1/D^2$ with

Margin notes (handwritten):

celestial objects emit EM radiation – poss to collect the spectral distribution

↓

recognising the spectral signature of an accretion disc if you saw one

|

Can calculate the emitted spectrum

① in equilib, the flux emerging from the disc annulus = rate at which viscous dissipation deposits energy into the disc ring. (we calculated L. of disc ring)

② if the disc is optically thick (opaque) the flux emerging from the ring is that of a BB described by its surface T.

③ thus T rises towards the centre of the disc. – the resultant spectrum is a superposition of many Planck functions with different Ts.

Activity 23
FKR section 5.5 + first sentence on p. 92.

NOTE
double use of symbols
V, D

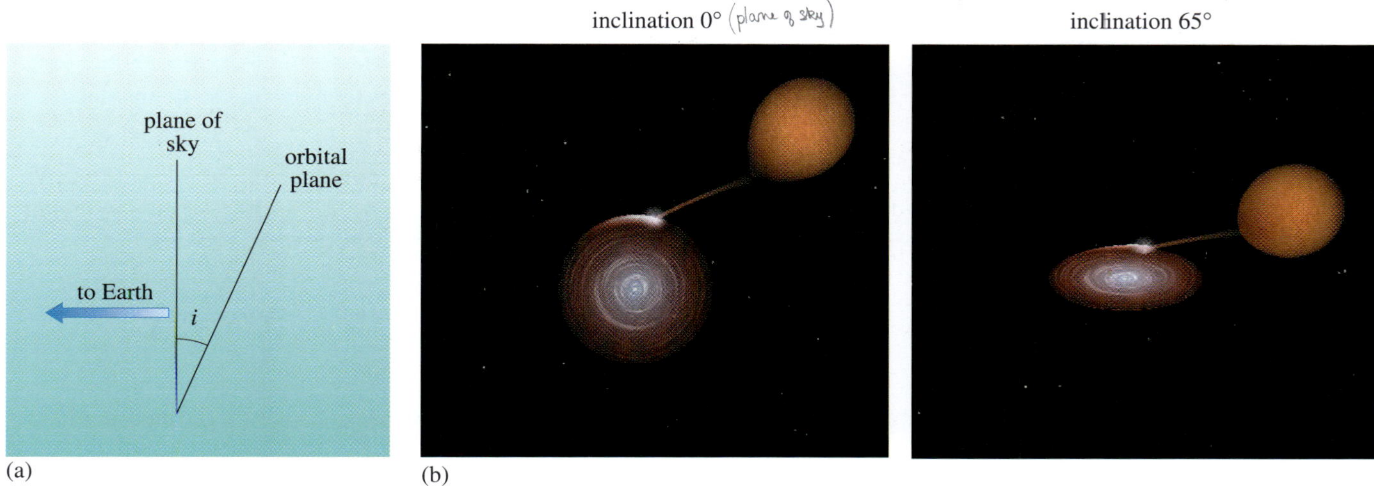

Figure 44 (a) Definition of the inclination i of a binary orbit. (b) A binary system viewed at two different inclination angles, $i = 0°$ and $i = 65°$.

distance D to the binary system: the total radiated energy is distributed over a larger and larger surface area $\propto D^2$ when it propagates away from the source. So the flux (energy per unit area) thins out. Formally, the flux F_ν is obtained by multiplying the intensity I_ν emitted by a disc ring with the solid angle subtended by the ring as seen from Earth. The solid angle is just the projected area of the disc ring, divided by D^2. The disc will in general be inclined to the line of sight. This is usually quantified in terms of the binary inclination i, the angle between the plane of the sky and the disc plane (see Figure 44). Hence the projected area of a disc ring at radius R with width $\mathrm{d}R$ is $2\pi R(\cos i)\mathrm{d}R$ (see Figure 45).

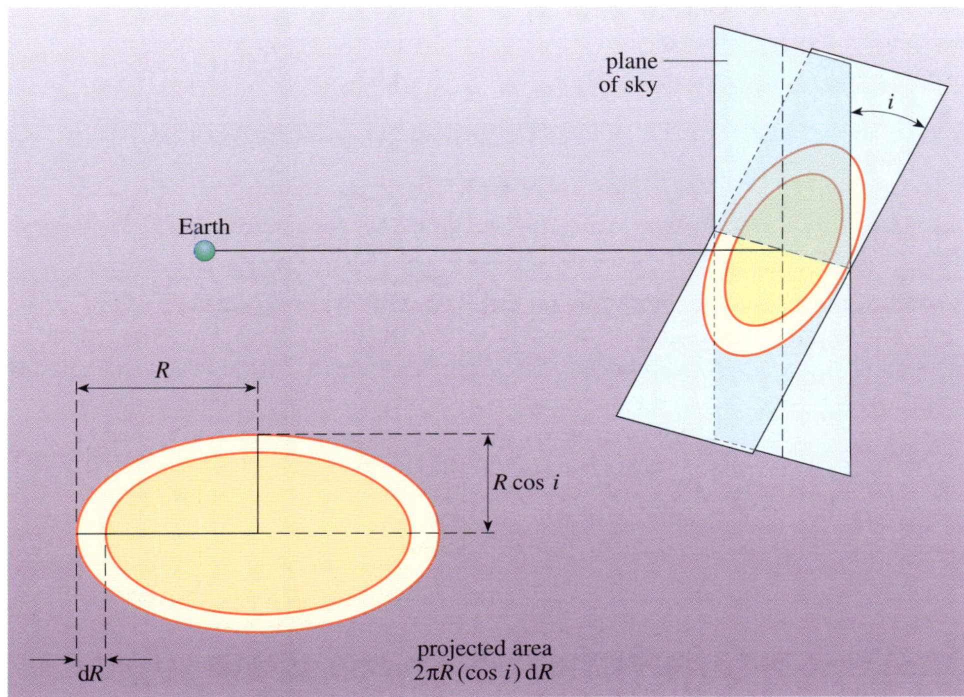

Figure 45 Solid angle subtended by a disc ring, as seen from Earth.

■ If the binary inclination is i, what is the angle between the line of sight and the orbital plane of the binary?

❑ The disc plane is identical to the orbital plane of the binary. Hence the angle between the orbital plane and the plane of the sky is just i, while the angle between the orbital plane and the line of sight is $90° - i$. ■

Question 39

Show that $T(R)$ as given by

$$T^4(R) = \frac{3GM\dot{M}}{8\pi R^3 \sigma}\left[1 - \left(\frac{R_*}{R}\right)^{1/2}\right] \qquad \text{(FKR 5.43) (65)}$$

attains a maximum value of $0.488T_*$ at $R = (49/36)R_*$, where T_* is defined in FKR Equation 5.44. *Hint*: Start from Equation 65. Then (a) substitute in T_*; (b) put $R/R_* = x$ and $(T/T_*)^4 = y$; (c) find dy/dx and set $dy/dx = 0$ to find the value of x where y is maximal; (d) find y at the maximum; (e) find T at the maximum. ■

Example 17

Why do we expect that an accretion disc around a white dwarf should emit primarily in the ultraviolet? Consider T_*.

Solution

The quantity T_* is a characteristic disc temperature (FKR Equation 5.44). For typical values of the mass transfer rate (= mass accretion rate), the mass of the white dwarf, and values of R_*, this temperature is of order 4×10^4 K (see FKR Equation 5.44, second line).

According to the Wien displacement law, the maximum of the Planck function B_λ characterizing a blackbody emission with temperature T is at a wavelength λ_{max}, given by

$$\lambda_{\text{max}} T = 2.9 \times 10^{-3}\ \text{m K}$$

(see Block 1, Section 4.9.3, Equation 156). Hence for T_* we have

$$\lambda_{\text{max}} = \frac{2.9 \times 10^{-3}\ \text{m K}}{T_*} = \frac{2.9 \times 10^{-3}\ \text{m K}}{4 \times 10^4\ \text{K}} = 7.25 \times 10^{-8}\ \text{m}$$

This is just longer than the wavelength that separates the X-ray and UV range, see the wavelength scale in Figure 9. ■

Question 40

Using the same arguments as in Example 17, show that an accretion disc around a neutron star emits predominantly in the X-ray region. ■

The main information you should take away from this section is summarized in FKR Figure 5.2. The integrated disc spectrum shown there is a 'stretched out' black body, with a rather *flat* part characterized by $F_\nu \propto \nu^{1/3}$. (You might think '*Not only is the disc flat, but even the spectrum is flat …*'.) Remarkably, the disc spectrum is independent of the viscosity in the disc!

The form $F_\nu \propto \nu^{1/3}$ is sometimes considered as a characteristic disc spectrum, but the width of this part of the spectrum depends on the temperature difference between the inner and outer disc edge. The outer edge of the disc is essentially fixed by the size of the binary, as the disc has to fit inside the primary star's Roche lobe. The inner edge of the disc depends on the radius of the central star. If this is a white dwarf the range in radii of the disc, and hence the temperature difference can in fact be too small to make the flat part of the spectrum noticeable. However, if the mass accreting star is a neutron star or black hole the inner disc radius can easily be three orders of magnitude smaller than for a white dwarf. The complication in this case is that the inner disc becomes so hot that the outer parts of the disc are heated by absorbing electromagnetic radiation emitted by the inner parts. The temperature distribution and emitted spectrum of such *irradiated* discs is different from what we have considered here. (This is covered in FKR Section 5.10, which is not part of this course.) Although more convincing evidence for the existence of accretion discs must await our discussion of the line spectrum below, we nonetheless take a look at an actually observed continuum spectrum of an X-ray binary.

Activity 24 (1 hour)

Nova Muscae 1991

Connect to the S381 home page, select Block 3, Activity 24 and follow the instructions on screen on how to download the paper entitled 'The Hubble Space Telescope observations of X-ray Nova Muscae 1991 and its spectral evolution' by F. H. Cheng *et al*. in *The Astrophysical Journal*, Vol. 397 (1992), pages 664–673.

The article reports observations of the outburst of a particular X-ray nova (soft X-ray transient). The text is concise and full of facts and details; it is written for experts working in the field, certainly not meant for easy digestion. Bearing this in mind, read the Abstract, the Introduction and Section 4.1 of the paper. In particular, look at Figures 4 and 5. Try to understand the overall context, and avoid getting bogged down by details or incomprehensible technicalities. Also, glance at Section 3, but do not try to derive the equations given there.

Keywords: none ■

In Section 3 of the paper, do you recognize the equations describing the continuum emission of accretion discs? They are just alternative forms of the equations we have been discussing above. The one exception is the temperature law (Equation 1 in the paper). The radial dependence is a rather complicated function. The reason why the law is different from FKR Equation 5.43 is that the inner boundary condition is slightly different for the case of very compact neutron star or black hole accretors than for the more moderate case discussed by FKR. As a result, the integration constant in Equation 57 will be slightly different, giving an apparently different law. But in fact, it is not all that different.

Question 41

Show that for large radii R the temperature law given by Cheng *et al*. (their Equation 1) is the same as the one given by Equation 65. (*Hint*: Show that both temperature laws give the same expression for T^4 in the limit $R \rightarrow \infty$.) ■

Question 42

Consider now Figures 4 and 5 of the paper, and try to answer the following questions.

(a) Which parts of the electromagnetic spectrum does the frequency range in the two figures cover (compare with Figure 9 in this Study Guide)? In what part of the electromagnetic spectrum are the observed data?

(b) What quantity is plotted against frequency?

(c) What time period do the IUE and HST spectra in Figure 5 cover?

(d) Which of the spectra shown in Figure 5 were taken by the Hubble Space Telescope? How long after the outburst of Nova Muscae have they been obtained?

(e) The figures compare the observed spectra with calculated spectra of steady-state accretion discs. What parameter has been varied to generate the array of theoretical spectra shown in the figures? *Variable parameter is the mass accretion rate Ṁ*

(f) What seems to be happening to the accretion disc of Nova Muscae over the period covered by the observations? ■

if apparently the Ṁ decreases with t as do the disc surface ρ + T. With decreasing surface the total mass in the disc must decreased as well — disc drains into B.H.

Some of the data shown in the paper by Cheng *et al.*, the subject of Activity 24, were obtained by the International Ultraviolet Explorer (IUE), a satellite observatory which no longer exists. Observations made by IUE and most other satellite observatories are archived and accessible via the Internet. The next activity gives an example.

Activity 25
Bk 3 Activity 25
S381 home page.

Activity 25 (30 minutes)

A satellite data archive

Connect to the S381 home page, select Block 3, Activity 25 and follow the instructions on screen.

Keywords: none ■

Question 43

(43) UV light is absorbed by the Earth atmosphere

Both HST and IUE are satellite observatories. Why didn't Cheng *et al.* use ground-based telescopes to make their observations? ■

5.2 Flat stars: the vertical disc structure

So far we have been mostly concerned with the radial structure of the disc. Other than saying that the disc is geometrically thin, and that we use vertically integrated quantities we didn't bother too much about the detailed vertical disc structure. But it is exactly in this vertical structure where the similarity between discs and stars becomes most obvious, as we will discover in this section.

We begin by investigating under what conditions the disc really is geometrically thin.

5.2.1 Hydrostatic equilibrium

Activity 26

Activity 26
FKR Section 5.3

hydrostatic equilib (HE)
a situation in which forces acting on a fluid (normally gravitational force) are balanced by internal P of fluid (thermal, degeneracy, radiation P) so the fluid neither collapses or expands
Equation for HE for a spherical symmetrical self gravitation system: dP/dr = −GM(r)ρ/r²
if fluid inside a gravitational potential well that is not in HE will collapse on a free fall timescale . until it reach in HE, Stars are close to HE most of time

Thin disc condition

You should now resume reading Section 5.3 in FKR on page 87. Start where it says 'We are now in a position to check some of the assumptions ...', and stop at the bottom of this page, where the text refers to later chapters ('In Section 5.6 we shall show ...').

Keywords: **hydrostatic equilibrium** ■

The starting point is once more the Euler equation, the equation of motion for fluid flow. As we are considering the vertical structure we are interested in forces acting in the direction perpendicular to the disc plane, the *z*-direction. In hydrostatic equilibrium the z-component of the gravitational attraction towards the central body must be balanced by a pressure gradient in the z-direction. The equation preceding FKR Equation 5.24, the z-component of the Euler equation, expresses just that. The acceleration due to the vertical component of the pressure gradient is the familiar term $\rho^{-1}(\partial P/\partial z)$ (see Equation 7), while the gravitational acceleration is calculated as the z-component of the gradient of the gravitational potential

$$V(r) = -\frac{GM}{r}$$

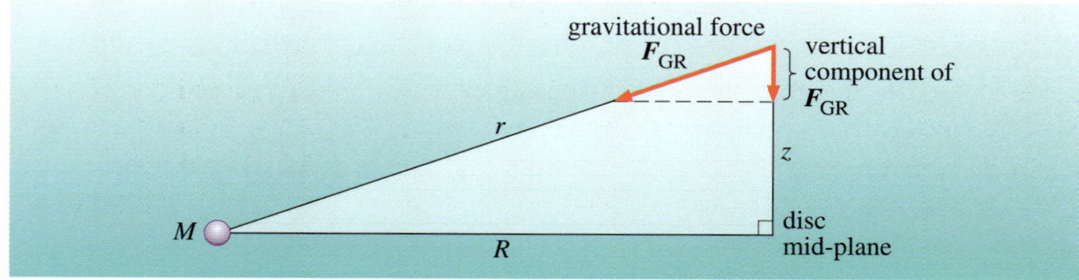

Figure 46 Component of gravity perpendicular to the disc plane.

The distance r to the central body of mass M (see Figure 46) can be written in cylindrical coordinates (z, R) as

$$r = \sqrt{R^2 + z^2}$$

Hence the acceleration g_z in the z-direction is

$$g_z = -\frac{\partial}{\partial z} V(R, z) = \frac{\partial}{\partial z} \frac{GM}{(R^2 + z^2)^{1/2}}$$

To carry out the differentiation we apply the chain rule on the variable $u = R^2 + z^2$.

$$-\frac{\partial}{\partial z} V(R, z) = \frac{\partial}{\partial z} \frac{GM}{u^{1/2}} = \left(\frac{\partial}{\partial u} \frac{GM}{u^{1/2}}\right) \frac{\partial u}{\partial z} = \left(\frac{\partial}{\partial u} \frac{GM}{u^{1/2}}\right) \frac{\partial (R^2 + z^2)}{\partial z}$$

Carrying out the two derivatives this becomes

$$-\frac{\partial}{\partial z} V(R, z) = \left(-\frac{1}{2} \frac{GM}{u^{3/2}}\right) 2z$$

Substituting again for u and then factoring out R^3 in the denominator gives

$$-\frac{\partial}{\partial z} V(R, z) = -\frac{GM}{(R^2 + z^2)^{3/2}} z = -\frac{GMz}{R^3[1 + (z/R)^2]^{3/2}}$$

For $z \ll R$ the term in square brackets can be taken to be unity, and we are left with what is written on the right-hand side of FKR Equation 5.24,

$$g_z = -\frac{GMz}{R^3} \qquad (66)$$

FKR shows that the Euler equation, FKR Equation 5.24, describing hydrostatic equilibrium in the vertical direction is equivalent to the following relation between the disc thickness (or scale height) H and the radius R.

$$H \cong \frac{c_s}{v_\phi} R \qquad (67)$$

This can also be written as $H/R \cong c_s/v_\phi$. In words: the disc scale height relates to the distance from the rotational axis as the sound speed to the azimuthal speed (i.e. in practice the Keplerian speed). An alternative way to derive Equation 67 is considered in Question 45, after the next reading.

Question 44

Show that the condition for a disc to be thin is really a condition on the disc temperature. Consult Equation 67 and the expression for the sound speed (Equation 26). ■

- Using the α-viscosity, show that in a thin disc the radial drift velocity is highly subsonic.

- From Equation 54 we have $v_R \sim v/R$. The α-viscosity takes the form $v = \alpha c_s H$ (Equation 39). Hence $v_R \sim \alpha(H/R)c_s$. But as $H/R \ll 1$ by assumption and $\alpha < 1$, therefore $v_R \ll c_s$. ■

This is an important result. From Equation 67 we see that in a thin accretion disc $(H \ll R)$ the azimuthal motion v_ϕ of the disc plasma on Keplerian orbits is highly supersonic ($c_s \ll v_\phi$), while the radial drift velocity is highly subsonic:

$$v_R \ll c_s \ll v_\phi \qquad (68)$$

We are now ready to read about flat stars.

5.2.2 Density structure and energy transport

ROSSELAND MEAN OPACITY

Opacity measures the opaqueness of matter against radiation. Specifically, κ is the absorption cross-section per unit mass, e.g. in $cm^2\,g^{-1}$. The cross-section and hence the opacity is a complicated function of the frequency of radiation. In the study of the stellar interior, and also in the context of the vertical structure of optically thick discs, we are not interested in this detailed frequency dependence. We work with a suitable average of the opacity, the Rosseland mean, usually denoted by κ_R. The Rosseland mean allows one to calculate the effect of energy transport by radiation in local thermodynamic equilibrium in the simple radiative diffusion approximation (FKR Equation 5.37).

(handwritten) $F(z) = \dfrac{-16\,\sigma T^3}{3\,\kappa_R \rho} \cdot \dfrac{\partial T}{\partial z}$ --- 5.37

Activity 27 (20 minutes)

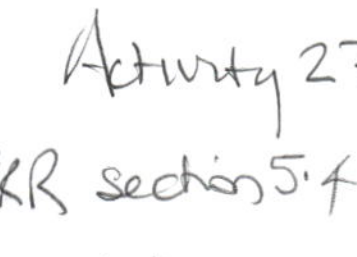

The local structure of thin discs

Read Section 5.4 of FKR. (Note τ is the optical depth.)

Keywords: **Rosseland mean opacity, optical depth** ■

FKR operates here with an isothermal vertical disc structure, i.e. FKR assumes that in a given disc ring the gas temperature is the same for all heights z. Clearly this cannot be the case exactly. Without a temperature gradient in the vertical direction the disc could not cool by radiation. But it turns out that in realistic disc models the temperature does not change a great deal from mid-plane to surface, perhaps only by

(page number, printed in body) 96

(handwritten margin notes, left side)

44

$H \sim R \dfrac{c_s}{v_\phi}$ — 67) ...(67)

$c_s \sim 10 \left[\dfrac{T}{10^4 K}\right]^{1/2}$... (26)

$H \sim R \cdot 10 \dfrac{\left[\dfrac{T}{10^4 K}\right]}{v_{10}}$

For disc to be thin $H \lll R$

Write (67) as $\dfrac{c_s R}{v_\phi} \lll R$

$\therefore c_s \lll v_\phi$

From 26 $c_s \propto T^{1/2}$

if c_s is required to be small so must T

(handwritten, lower left)

Activity 27

FKR section 5.4

optical depth – quantity used to describe the transparency of a region of space containing obscuring material to EMR of a certain λ. If a region of space has an optical depth of τ at a given λ then the measured flux of light that has passed thro that region is reduced by a factor $\exp(-\tau)$ cpd with what would be expected if there was no obscuration. medium is transparent if $\tau \approx 0$, if $\tau \lll 1$ medium is optically thin, if $\tau \ggg 1$ medium is optically thick. The optical depth can be related by opacity $\tau = \int \kappa(z) \rho(z) dz$, ($\kappa$ opacity at

a factor of a few, unlike in the radial direction where the temperature changes by orders of magnitude. So the assumption of an isothermal vertical disc layer is not all that bad.

Question 45

Show that in the case of constant temperature in z-direction, the Euler relation FKR Equation 5.24 can be rearranged to give the condition on H/R expressed in Equation 67. (*Hint*: Insert the density law FKR Equation 5.33 into FKR Equation 5.24, and use Equation 25 for the isothermal sound speed to replace pressure with density.) ■

Example 18

Calculate how the disc surface density Σ relates to the mid-plane density ρ_c if the vertical density law FKR Equation 5.33 is assumed.

Solution

In the direction perpendicular to the disc plane the mass density drops off as (FKR Equation 5.33 and Figure 47)

$$\rho(z) = \rho_c \exp\left(-\frac{z^2}{2H^2}\right)$$

The surface density (Equation 33) is defined as the integral

$$\Sigma = \int_{-\infty}^{+\infty} \rho(z)\,\mathrm{d}z = \rho_c \int_{-\infty}^{+\infty} \exp\left(-\frac{z^2}{2H^2}\right)\mathrm{d}z$$

There is a standard integral which has a form very similar to this:

$$\int_0^{\infty} \exp(-a^2x^2)\,\mathrm{d}x = \frac{\sqrt{\pi}}{2a}$$

(*Standard* means that this is a well-known result from mathematics; should you ever need a standard integral you will either be told what it is or where to look it up.) The disc is symmetrical to the mid-plane, so we write

$$\Sigma = 2\rho_c \int_0^{+\infty} \exp\left(-\frac{z^2}{2H^2}\right)\mathrm{d}z$$

Then we substitute $a^2 = 1/(2H^2)$ to get

$$\Sigma = 2\rho_c \int_0^{+\infty} \exp(-a^2z^2)\,\mathrm{d}z$$

hence from the standard integral

$$\Sigma = 2\rho_c \frac{\sqrt{\pi}}{2a}$$

So, substituting back for a,

$$\Sigma = 2\rho_c \frac{\sqrt{\pi}}{2\sqrt{1/2H^2}} = \sqrt{2\pi}\rho_c H$$

The numerical factor $\sqrt{2\pi}$ is of order unity (well – almost). Real discs are not isothermal and the density law is somewhat different, so a slightly different numerical factor would arise.

This is a slightly more refined version of the one-zone model of vertical disc structure, where relations of the type $\Sigma = H\rho$ are used.

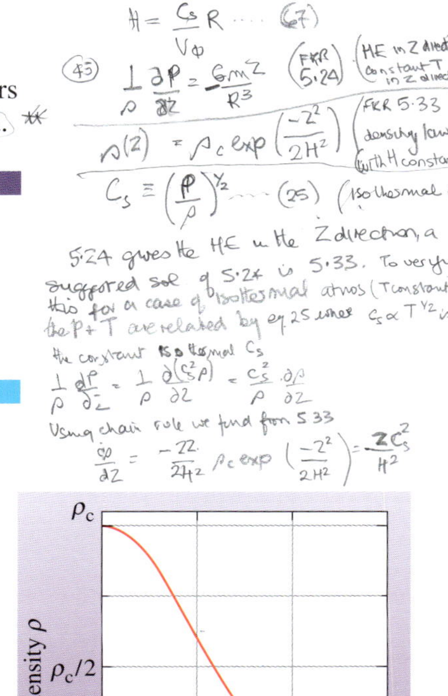

Figure 47 Vertical density profile for an isothermal disc.

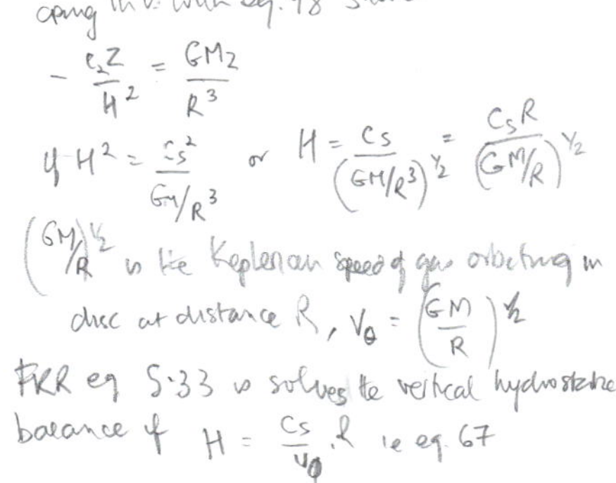

In the previous section, in the context of the radial disc structure, we talked at length about the conservation of mass and angular momentum. It is only now that we make explicit use of the other fundamental conservation law in classical physics, the **conservation of energy**. Viscous dissipation deposits energy locally in each disc ring and heats it up. FKR subsumes this in the symbol Q^+, the rate of energy produced per unit volume. At the same time energy is lost from the disc ring. Radiation (and perhaps convection) moves it away from where it has been generated, towards the surface of the disc. The flux F – not to be confused with F_ν used in Section 5.1 – measures the energy flow per unit time and unit surface area perpendicular to the disc plane. In equilibrium, the flux leaving a volume element upwards must be larger than the flux entering it from below if energy is generated in this volume element (Figure 48). This is expressed in the energy balance equation

$$\frac{\partial F}{\partial z} = Q^+$$

which is reminiscent of the stellar structure equation describing the nuclear energy generation in stars,

$$\frac{\partial L}{\partial r} = 4\pi r^2 \varepsilon$$

which you should have seen before (e.g. Phillips Equation 3.26).

■ Explain the difference between the symbols used in the above energy equations for stars and discs.

❑ The luminosity L is the energy flowing through a sphere with surface $4\pi r^2$ per unit time. Hence the flux is $F = L/4\pi r^2$. In stars the energy flows in a radial direction, so $\partial/\partial r$ takes the place of $\partial/\partial z$. In stars the energy generation rate per unit volume, ε, is from nuclear reactions, while in discs the energy generation rate per unit volume, Q^+, is from viscous dissipation. ■

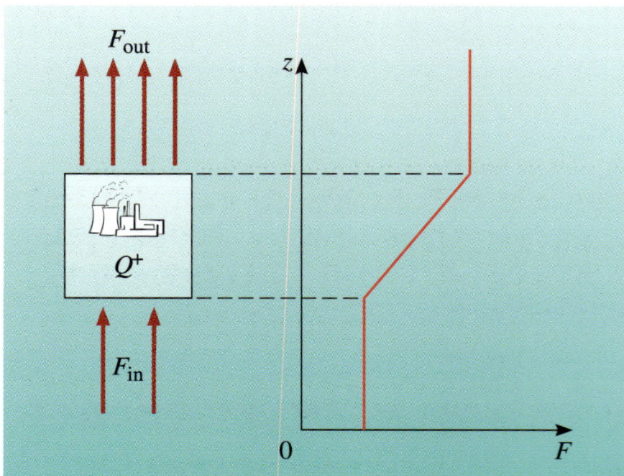

Figure 48 Balance of heating and cooling in the z-direction.

The optical depth τ is defined as the integral of the volume density of the absorbing cross-section, $\rho\kappa$, along the path the radiation has to travel. In the vertical direction, the optical depth of a layer at height z above the disc mid-plane is

$$\tau(z) = \int_z^\infty \rho(z')\kappa_R(z')\,dz'$$

Note that we adopt the symbol z' for the integration variable, as z is already used to denote the level in the disc whose optical depth we wish to calculate.

Question 46

Show that the optical depth is dimensionless. ■

The optical depth to the disc mid-plane is $\tau(0)$. This can be written as

$$\tau(0) = \int_0^\infty \rho(z')\kappa_R(z')\,dz'$$

Introducing the vertical average $\langle \kappa_R \rangle$ this becomes

$$\tau(0) = \langle \kappa_R \rangle \int_0^\infty \rho(z')\,dz'$$

$\langle \kappa_R \rangle$ is in fact defined by this relation. As the disc is symmetric to the mid-plane the remaining integral can also be written as (using Equation 33)

$$\int_0^{+\infty} \rho(z')\,dz' = \frac{1}{2}\int_{-\infty}^{+\infty} \rho(z')\,dz' = \frac{\Sigma}{2}$$

so that the optical depth finally becomes

$$\tau(0) = \langle \kappa_R \rangle \frac{\Sigma}{2}$$

FKR uses the typical optical depth

$$\tau = \Sigma\langle \kappa_R \rangle \tag{69}$$

simply written as $\tau = \Sigma\kappa_R$, to characterize whether the disc is opaque ($\tau \gg 1$) or not.

If the energy is carried away by radiation, i.e. if the disc is **radiative**, the flux is given by FKR Equation 5.37. Compare this to the corresponding stellar structure equation in Block 2 (Section 2.2.3, Equation 6). FKR simplifies the energy balance condition considerably by replacing the gradient with an average, and by noting that $T_c^4 \gg T_{\text{eff}}^4$. We shall revisit the energy balance in some detail in Section 8, so if the simplifications worry you here, perhaps it will all become more transparent in Section 8.

Question 47

Consider FKR Equation 5.41, the collected set of equations describing thin discs. For each of the eight equations in the set, describe what they represent in physical terms. ■

5.3 Shakura–Sunyaev discs

By now we already have a fairly complete quantitative description of a steady-state accretion disc. At this point it is instructive to see the disc equations 'live', to fill them with actual numbers. You will gain a feeling for how large and hot typical discs are, and how much mass they contain. In order to do so we have to specify the **input physics** – the equation of state, the opacity and viscosity – i.e. the functions that describe the property of stellar plasma in general.

In what follows we adopt the α-viscosity introduced in Section 3 above. Discs subject to this form of viscosity are sometimes called Shakura–Sunyaev discs, in honour of Nikolai Shakura and Rashid Sunyaev who pioneered the study of α-discs in 1973. The next activity introduces the full set of equations describing Shakura–Sunyaev discs.

Activity 28 **(30 minutes)**

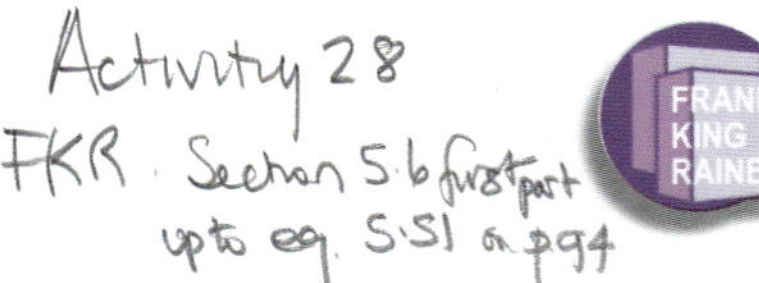

α-discs

You should now read the first part of Section 5.6 of FKR up to and including the first sentence after FKR Equation 5.51 on page 94.

Do not try to follow all the algebraic manipulations described in this section of the book. Rather, you should understand where each of the equations in the full set FKR Equation 5.49 comes from, and look at the magnitude and radial dependence of the different quantities.

Keywords: **Shakura–Sunyaev discs, Kramers law** ■

You will have certainly noticed that most of the quantities that enter the equations in Section 5.6 of FKR are normalized to certain *typical* units. For instance, the mass accretion rate is given in units of $10^{16}\,\mathrm{g\,s^{-1}}$, i.e.

$$\dot{M}_{16} \equiv \frac{\dot{M}}{10^{16}\,\mathrm{g\,s^{-1}}}$$

simply because $10^{16}\,\mathrm{g\,s^{-1}}$ is a value one would typically expect for the mass transfer rate the secondary star of a cataclysmic variable supplies into the disc. As the disc is in a steady state by assumption, we know that the local mass flow rate through the disc must be equal to the mass transfer rate at all radii.

The radius variable R is given in units of 10^{10} cm, because this is the typical size of a disc in a CV. Likewise, the mass of the accreting star is in units of $1\mathrm{M}_\odot$, representative of the mass of a white dwarf. The advantage of this is obvious. The numerical coefficients that appear, e.g. in the individual equations of the system FKR Equation 5.49, immediately indicate the order of magnitude we have to expect for the various disc quantities. This is because the remaining factors are all of order unity. So by just looking at, say, the equation for the radial drift velocity, we immediately see that v_R is of order $10^4\,\mathrm{cm\,s^{-1}}$, i.e. almost $400\,\mathrm{km\,h^{-1}}$, the speed of a high-speed train (not in the UK, though ...). This sounds fast, but keep in mind that the material has to 'drift' inwards through the entire disc, i.e. for about 10^{10} cm, (i.e. 100 000 km!). This therefore takes of order 10^6 s, or about one month. In fact, we have just estimated the typical viscous timescale of such a disc if $\alpha \approx 1$.

Question 48

Why is it not possible to discover the magnitude of viscosity (the value of α) by just observing a steady-state disc? ■

The drift velocity we have just calculated still sounds fast. Is it really true that it is negligible compared to the azimuthal velocity of disc material?

Example 19

Compute the azimuthal velocity of plasma in a typical CV accretion disc, at a radius $R_{10} = 1$ and $R_{10} = 0.05$.

Solution

The disc plasma follows Kepler orbits. Hence the azimuthal or orbital velocity v_ϕ is just the Keplerian velocity v_K as given by our Equation 1,

$$v_\mathrm{K} = \sqrt{\frac{GM}{R}}$$

We rewrite this in terms of the dimensionless radius R_{10} and mass m_1

$$v_\mathrm{K} = \sqrt{\frac{GM_\odot m_1}{10^{10}\,\mathrm{cm}\,R_{10}}} = \sqrt{\frac{6.673\times10^{-11}\,\mathrm{N\,m^2\,kg^{-2}}\times1.99\times10^{30}\,\mathrm{kg}}{10^8\,\mathrm{m}}}\sqrt{\frac{m_1}{R_{10}}}$$

$$= 1.2\times10^6\,\mathrm{m\,s^{-1}}\sqrt{\frac{m_1}{R_{10}}}$$

So for a typical mass, $m_1 = 1$, we have at $R_{10} = 1$

$$v_\mathrm{K} = 1.2\times10^6\,\mathrm{m\,s^{-1}}\sqrt{\frac{1}{1}} = 1.2\times10^6\,\mathrm{m\,s^{-1}} = 1.2\times10^8\,\mathrm{cm\,s^{-1}} \approx 10^4 v_\mathrm{R}$$

and at $R_{10} = 0.05$

$$v_\mathrm{K} = 1.2\times10^6\,\mathrm{m\,s^{-1}}\sqrt{\frac{1}{0.05}} = 5.4\times10^6\,\mathrm{m\,s^{-1}} = 5.4\times10^8\,\mathrm{cm\,s^{-1}}$$

$$\approx 5\times10^4 v_\mathrm{R}$$

The azimuthal velocity is indeed more than 10 000 times larger than the radial drift velocity $v_\mathrm{R} = 10^4\,\mathrm{cm\,s^{-1}}$, thus confirming Equation 68. ■

Question 49

Using the set FKR Equation 5.49, derive FKR Equation 5.50 for the disc thickness. ■

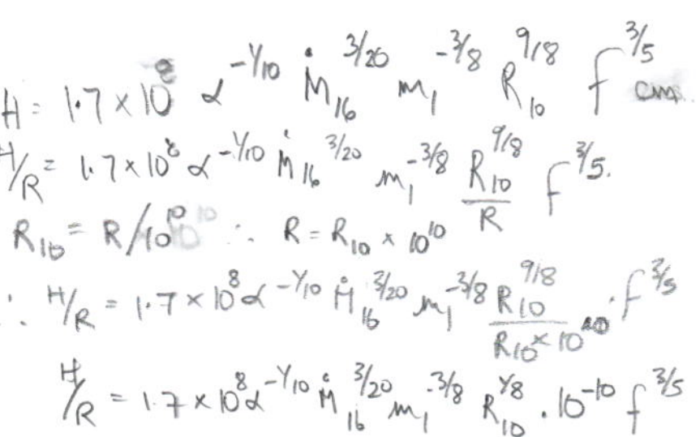

Question 50

Using the set FKR Equation 5.49, estimate the disc mass, i.e. derive FKR Equation 5.51. (*Hint*: FKR Equation 5.51 describes an upper limit for the disc mass. To obtain this limit, set $f = 1$. Also, assume that the outer disc radius is at 10^{11} cm, while the inner disc radius is 0.) ■

As it is so much fun finally to see the structure of a disc, let us plot it.

Activity 29 **(1 hour)**

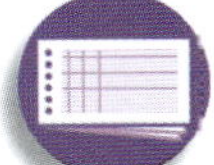

Radial profiles of α-discs

Use the set of equations in FKR Equation 5.49 and FKR Equation 5.50 to calculate and plot (a) the surface density profile $\Sigma(R)$, (b) the relative disc thickness H/R and (c) the disc mid-plane temperature $T(R)$ as a function of distance R from the central accreting body.

Assume that the accretor is a white dwarf with mass $0.6\mathrm{M}_\odot$, and that the inner disc radius is $R_* = 10^9$ cm, while the outer disc radius is 10^{10} cm.

8.7 ×10^8 cms. The sample ss shows all functions out to radius 10^{11} cms

Plot the profiles for three different values of the mass transfer rate (e.g. $10^{-10}\mathrm{M}_\odot\,\mathrm{yr}^{-1}$, $10^{-9}\mathrm{M}_\odot\,\mathrm{yr}^{-1}$, and $10^{-8}\mathrm{M}_\odot\,\mathrm{yr}^{-1}$) and for $\alpha = 0.3$.

As a first step you will need to convert these mass transfer rates into cgs units.

If you have difficulties with this activity, refer to the ‘Comments on activities’ section.

We shall come back to these surface density profiles in the main multimedia tutorial, Activity 48.

Keywords: none ■

The surface density profiles will play a vital role in our interpretation of the dwarf nova and soft X-ray transient phenomena. So we will consider them further here. To calculate the vertical structure of the disc we made the assumption that energy transport is via radiation, i.e. through the slow diffusion of photons to the surface. This process is controlled by the opacity κ. In particular, in order to maintain a fixed energy flux F the required temperature gradient $\partial T/\partial z$ is proportional to κ, see FKR Equation 5.37.

■ In the set of equations FKR 5.41, which equation(s) refer to radiative energy transport in the z-direction?

❑ It is Equation no. 5. The left-hand side expresses the energy that is transported away per unit time via radiative transfer in the vertical direction, i.e. from the disc mid-plane to the surface of the disc. The right-hand side of the equation is just the viscous dissipation rate in the disc, i.e. the rate at which energy is deposited into the disc. Hence Equation 5 expresses the balance between viscous heating and radiative cooling. Equation 6 is also related to radiative energy transport in z-direction: it defines the optical depth of the disc mid-plane, $\tau = \kappa\rho H = \kappa\Sigma$, which appears in Equation 5. ■

(5) $$\frac{4\sigma T_c^4}{3\tau} = \frac{3GM\dot{M}}{8\pi R^3}\left[1-\left(\frac{R_*}{R}\right)^{1/2}\right]$$

energy that is transported away per unit time via radiative transfer vertically (from disc midplane to surface of disc) — viscous dissipation rate in the disc (the rate at which energy is deposited into the disc)

eq 5 expresses the balance between viscous heating & radiative cooling

eq 6 $\tau = \Sigma \kappa_R(\rho, T_c) = \tau(\Sigma, \rho, T_c)$ is related to radiative transport in Z direction because it defines the optical depth of disc mid plane, $\tau = \kappa\rho H = \kappa\Sigma$. τ appears in eq. 5

ENERGY TRANSPORT BY CONVECTION: A SUMMARY

In convective layers most of the energy is effectively trapped and transported in rising blobs of gas. These **convective eddies** usually do not exchange energy with the surrounding medium while they travel, so they behave effectively adiabatically. Once they have risen for about a pressure scale height (the height over which pressure drops by a factor e) they finally mix with the surrounding medium, depositing their excess energy at this point. As the energy is literally moving *with* the matter, convection is usually a very effective means of energy transport. For it to work there must be an upward force or buoyancy on the eddies. This occurs *only* if the drop in temperature in the surrounding medium is *steeper* than the drop in temperature in the eddy itself. For if this is the case the eddy will be hotter than the surrounding medium after it has risen a little. Consequently it will have expanded somewhat, so that its density is smaller than in the surrounding medium. This causes the desired upward lift by buoyancy. If effective convection occurs, it sets up the adiabatic temperature gradient (i.e. the temperature gradient under adiabatic conditions; see Section 5.5.1 of Block 2) in the disc layers.

As you will recall from Block 2 (Section 5.5.1), energy transport can also occur in the form of **convection**. Convection occurs when the vertical temperature gradient that the disc would adopt in the absence of convection is too steep, i.e. steeper than the critical temperature gradient under adiabatic conditions. A prime cause for such a steep temperature gradient is an unusually large opacity. The opacity does become particularly large when hydrogen, the dominant species in the accretion disc plasma, starts to recombine, i.e. when the ionization of hydrogen is incomplete. Then bound-free and bound-bound transitions become available, in addition to the free-free transitions which dominate the opacity in a fully ionized gas. The opacity no longer follows Kramers law.

■ What is Kramers law?

❑ Kramers law (Kramers opacity) describes the Rosseland mean opacity of a plasma with density ρ and temperature T as a simple power law, $\kappa_R \propto \rho T^{-3.5}$. This fit is a good approximation to the actual opacity when free-free and, to some extent, bound-free interactions dominate (see, for example, Block 2, Section 2.2.1). ■

Hydrogen is partially ionized in the temperature region around 6000–8000 K, so we expect convection to take over the vertical energy transport when the disc temperature reaches this value. Given the standard temperature profile $T \propto R^{-3/4}$ of a disc this will be quite far away from the central accreting star (see, for example, the temperature profiles you have plotted in Activity 29).

An important consequence of the increase in opacity and the onset of convection is that the surface density profile $\Sigma(R)$ changes as well. The standard form for radiative discs is $\Sigma(R) \propto R^{-3/4}$ (FKR Equation 5.49). The surface density decreases as we move away from the centre of the disc. But when convection sets in Σ starts to increase again, and remains roughly constant at yet larger distances (see Figure 49).

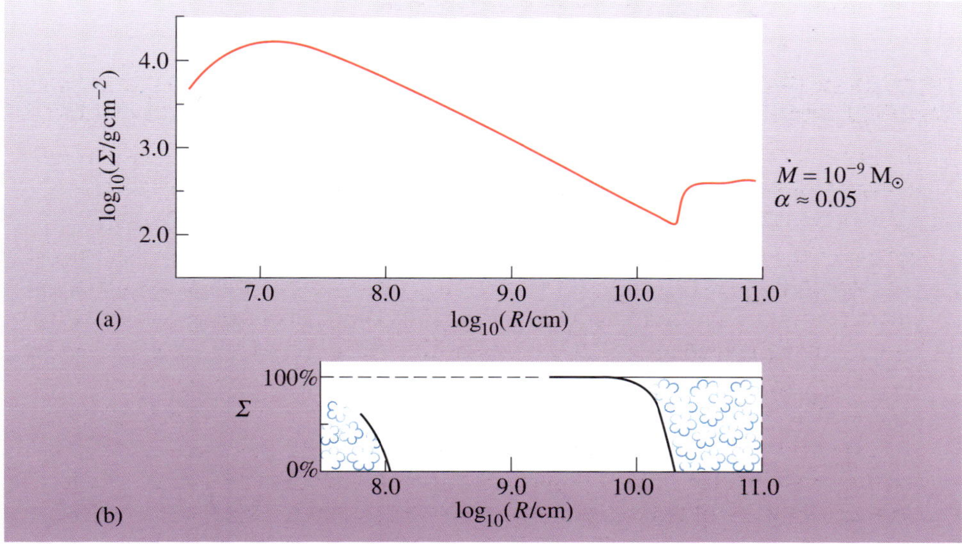

Figure 49 (a) Surface density profile of a stationary α-disc (accretor mass $1\mathrm{M}_\odot$, mass accretion rate $10^{-9}\mathrm{M}_\odot\,\mathrm{yr}^{-1}$, $\alpha \approx 0.05$). (b) Mode of vertical energy transport. 'Clouded' regions are convective.

As we shall see in Section 8, accretion discs that are large enough that they become convective in their outer parts are *unstable*, i.e. they cannot really exist in the hypothetical steady state we are assuming here. You will learn more about this later.

With convection, the disc equations are no longer as easy to solve as for radiative discs. We have to rely on numerical calculations to obtain a result like the one shown in Figure 49. But it is easy to see why the surface mass density in a convective disc must be larger than in a radiative disc. The magnitude of the temperature gradient established by convection is *smaller* than the magnitude of the temperature gradient in a radiative disc. (Note that the gradient is negative in both cases.) Hence for a given surface temperature (6000 K, say) the convective layer has a *smaller* mid-plane temperature T_c than the hypothetical radiative layer. The α-viscosity $\nu \propto c_s H$ (Equation 39) is also proportional to the temperature, $\nu \propto T_c$ (see Question 51). Hence with a smaller mid-plane temperature the viscosity is smaller as well, and with it the local drift velocity $\upsilon_R \propto \nu/R$ (Equation 54). As both the convective and hypothetical radiative layer have to support the same local mass accretion rate $\dot{M}(R) = 2\pi R\Sigma(-\upsilon_R)$ (Equation 42), the smaller radial drift velocity must be compensated by a larger surface density Σ. In other words, the slower drift leads to an accumulation of mass.

Question 51

Show that the α-viscosity implies that $\nu \propto T_c$. (*Hint*: Use Equation 67 to replace H by c_s, and note that $c_s \propto T^{1/2}$.) ■

The name game: part 5

Skim-read Section 5 and make a list of all symbols that denote *new* variables and *new* quantities, and note down what they mean. Flag up those symbols that have (unfortunately) multiple uses.

Once you have compiled your list and explained the meaning of all entries, compare it to the one given in Appendix A5 at the end of the Study Guide.

5.4 Summary of Section 5

1. The inclination of a binary orbit is the angle between the orbital plane of the binary and the plane of the sky.
2. The radial surface temperature profile $T(R)$ of an optically thick, steady-state disc is given by

$$T^4(R) = \frac{3GM\dot{M}}{8\pi R^3\sigma}\left[1-\left(\frac{R_*}{R}\right)^{1/2}\right]$$

3. Accretion discs in cataclysmic variables emit predominantly in the ultraviolet, discs in neutron star or black hole systems in the X-ray regime.
4. The continuum spectrum of geometrically thin, optically thick Keplerian steady-state accretion discs is a 'stretched out' black body. The existence of the characteristic flat part $F_\nu \propto \nu^{1/3}$ of the spectrum depends on the temperature difference between the inner and outer disc edge. In white dwarf systems it is hardly noticeable. In neutron star or black hole systems the inner disc can be so hot that it irradiates the outer disc.

5 Observations made by satellite observatories are archived and usually accessible via the Internet.

6 In hydrostatic equilibrium the vertical component of the gravitational attraction towards the central body must be balanced by the vertical pressure gradient.

7 Accretion discs are geometrically thin if the Keplerian motion is highly supersonic. The disc thickness (or scale height) H is

$$H \cong \frac{c_s}{v_\phi} R$$

8 The equation describing the conservation of energy is $\partial F/\partial z = Q^+$ and expresses the balance between heating Q^+ due to viscous dissipation and cooling due to radiation or convection.

9 The Rosseland mean opacity is a frequency-average of the opacity calculated in such a way that energy transport by radiation can be written as a diffusion equation.

10 A characteristic measure for the optical depth of an accretion disc is the quantity $\tau = \Sigma\kappa_R$.

11 A Shakura–Sunyaev disc is a model of a geometrically thin, optically thick steady-state accretion disc with an α-viscosity.

12 If $\alpha \approx 1$ the radial drift velocity v_R is of order $10^4\,\mathrm{cm\,s^{-1}}$. The typical outer disc radius in short-period binaries such as CVs is of order $10^{10}\,\mathrm{cm}$, the relative thickness H/R is typically less than 1%. The mass in the disc is of order $10^{-10}\mathrm{M_\odot}$, the accretion rates between $10^{-10}\mathrm{M_\odot\,yr^{-1}}$ and $10^{-8}\mathrm{M_\odot\,yr^{-1}}$.

13 Discs become convective when the vertical temperature gradient the disc would adopt in the absence of convection is too steep. This occurs in particular when the ionization of hydrogen is incomplete.

14 The surface density in radiative discs drops as $R^{-3/4}$. At the point where convection sets in, Σ increases with R, and remains roughly constant at larger radii.

6 REALITY CHECK: THE TRUTH IS OUT THERE

At this point we have achieved a pretty detailed theoretical understanding of steady-state accretion discs. As you can attest, we followed well-founded physical principles and conservation laws to come up with a qualitative and quantitative description of an accretion disc. In fact, we have constructed a *model* for an accretion disc. But how do we know how good this model is? How accurately does it describe the accretion flow in actual binaries out there in the Universe? The only way to find out, and the ultimate test for *any* theoretical model of any physical system, is to check it against observations. Is there any evidence in support of our general picture? Are there any observations in conflict with the model?

The following section is a combination of various activities clustered around Section 5.7 of FKR. We have collected a number of figures and animation sequences that further illustrate the issues discussed in FKR. As we present these below in the order that they occur in the text of FKR you will end up jumping back and forth between the text, your computer, and this Study Guide. If you prefer to read the whole of Section 5.7 of FKR first (with the exception of pages 109 and 110) you are welcome to do so.

Activity 30

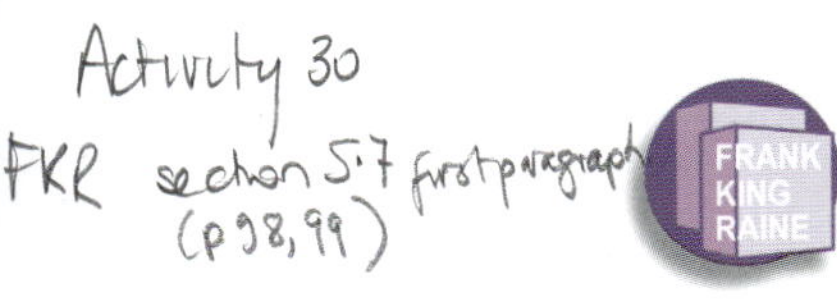

Compact binaries

Now read the first paragraph of Section 5.7 of FKR (on pages 98 and 99).

Keywords: none ■

You should remember low-mass X-ray binaries (LMXBs) and cataclysmic variables (CVs) from the previous sections, in particular Section 1. You have even seen images of how these systems might look (e.g. the cover image of this Study Guide, and Figure 14). A few technical terms used by FKR deserve a quick explanation.

Galactic bulge sources are steady, or mildly variable, X-ray sources located in the nuclear bulge of our Galaxy (see Block 1, Section 2.13). Their name goes back to the early days of X-ray astronomy when X-ray satellites for the first time established a group of bright X-ray sources in the Galactic bulge. Nowadays they are simply regarded as a subclass of LMXBs.

You have already encountered interacting binary stars such as dwarf novae and classical novae. FKR refers to classical novae as *old novae* (or post-novae), to specify the state of the system some time, perhaps years, after the actual nova outburst. The binary should have been much the same before the outburst, but usually the pre-novae are not known, and no previously known CV has ever been seen to undergo a nova outburst. In addition, and to make the confusion complete, there is a class of non-eruptive CVs, the **nova-like variables**, or simply *nova-likes*. They are similar to dwarf novae in outburst, but persistently bright.

In nova-likes and most dwarf novae, the white dwarf has either no magnetic field, or only a very weak magnetic field. But there are truly magnetic CVs as well. In *AM Herculis systems* (also often just called **AM Her stars**), named after the variable star AM in the constellation Hercules, the white dwarf is so strongly magnetic that the formation of an accretion disc is prevented altogether (see Figure 50). In

intermediate polars (IPs; sometimes also called **DQ Her stars**, after an atypical prototype) the field is weaker, or the binary separation larger, so that an accretion disc may exist, but its inner part, closest to the white dwarf where the magnetic effects are strongest, is perturbed or even disrupted (Figure 51). Should you wonder about the name, it might help to tell you that AM Her stars are also called **polars**, because a significant fraction of the light they emit is polarized. We shall hear more about polars and intermediate polars in Section 7.

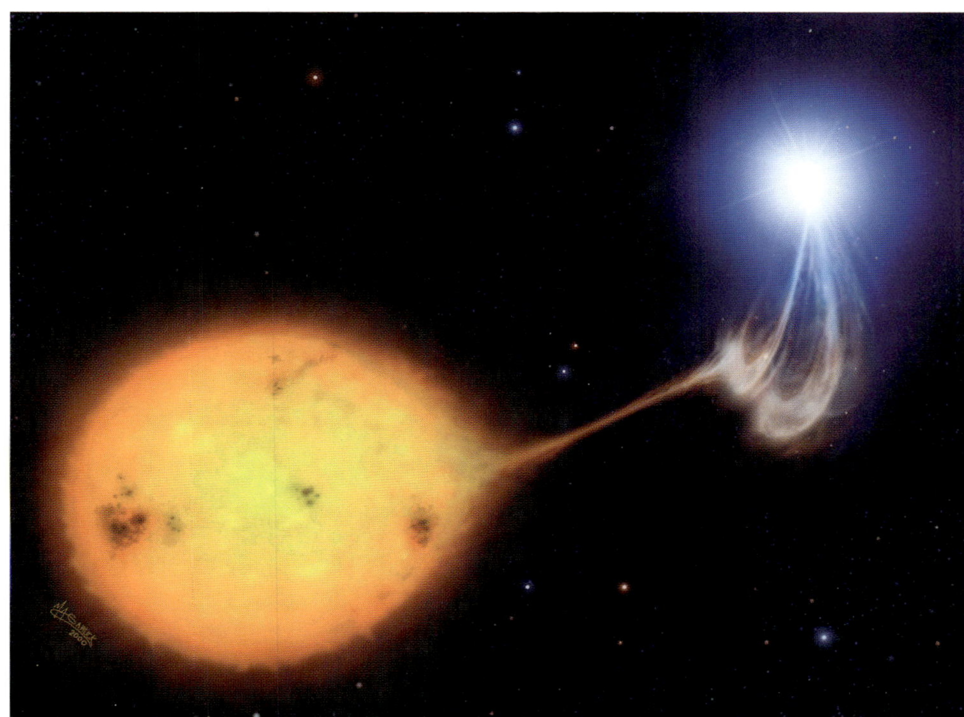

Figure 50 An artist's impression of a polar or AM Her star (strongly magnetic white dwarf).

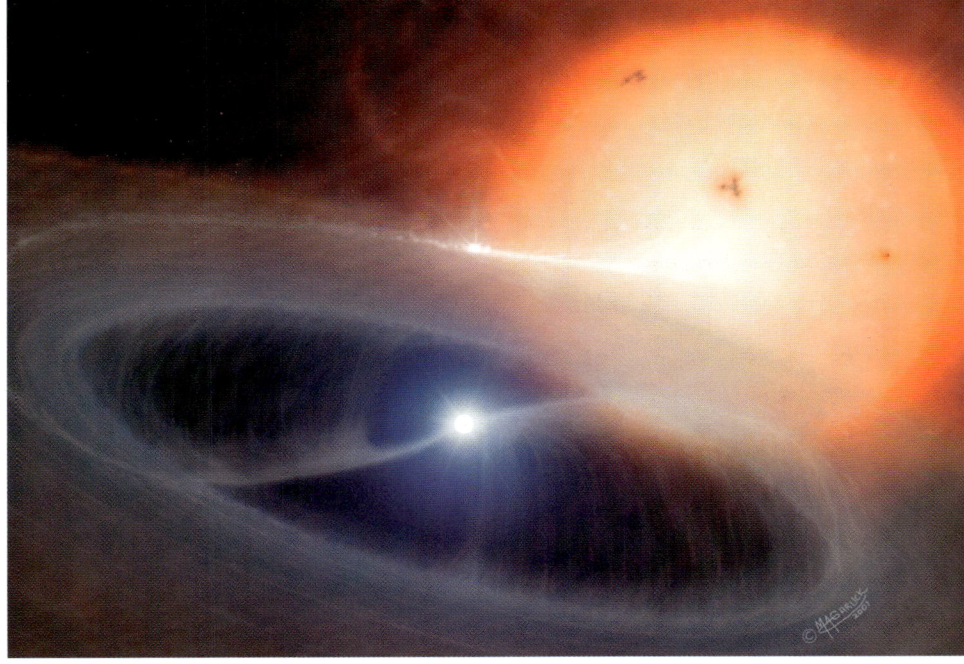

Figure 51 An artist's impression of an intermediate polar or DQ Her star (mildly magnetic white dwarf).

Question 52

Summarize the reasoning that led to the identification of CVs as the ideal host systems for the study of accretion discs. ■

Question 53

Most known CVs are fairly close, with typical distances in the range 100–200 pc. Here we consider the reason for this.

(a) Calculate the apparent visual magnitude of a typical CV located at a distance of 100 pc from the Sun, assuming that the light is dominated by accretion luminosity.

Hint 1: Calculate the accretion luminosity if the white dwarf has mass $1\mathrm{M}_\odot$ and radius $R = 8.7 \times 10^8$ cm, and the mass accretion rate is $10^{-9}\mathrm{M}_\odot\,\mathrm{yr}^{-1}$.

Hint 2: Calculate the absolute visual magnitude by assuming that the relation between the luminosity and the visual magnitude for this CV is the same as for the Sun. (The absolute visual magnitude of the Sun is $M_\mathrm{V} = 4.83$.)

Hint 3: Calculate the apparent visual magnitude, using the distance modulus.

(b) Why could this explain the fact that most CVs are fairly close? ■

You will remember that the continuum spectrum we observe from systems with a suspected accretion disc does not really provide a very convincing proof for the existence of a disc. Spectral lines are much better at this.

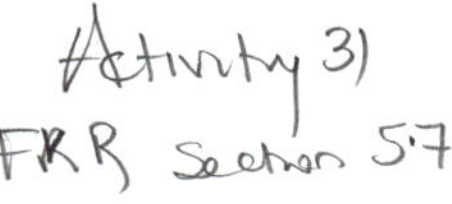

Activity 31

Double-peaked emission lines

Now read the second paragraph of Section 5.7 of FKR (pages 99 to 101). In parallel, have a look at our Figure 52 and Figure 53, and compare them with Fig. 5.5 in FKR.

Keywords: **double-peaked emission lines**, **eclipses** ■

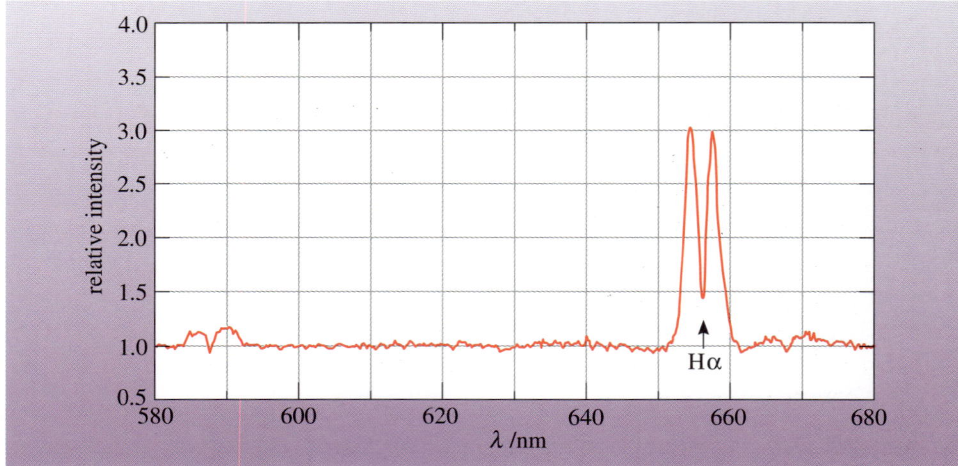

Figure 52 The optical spectrum of the cataclysmic variable WZ Sagittae in the wavelength range 580 nm to 680 nm. Wavelength is plotted along the *x*-axis, the intensity of the light at this wavelength is plotted along the *y*-axis. The prominent double-peaked feature labelled 'Hα' is the Hα emission line.

the dark green region should be the blue region
the blue region should be the dark green region

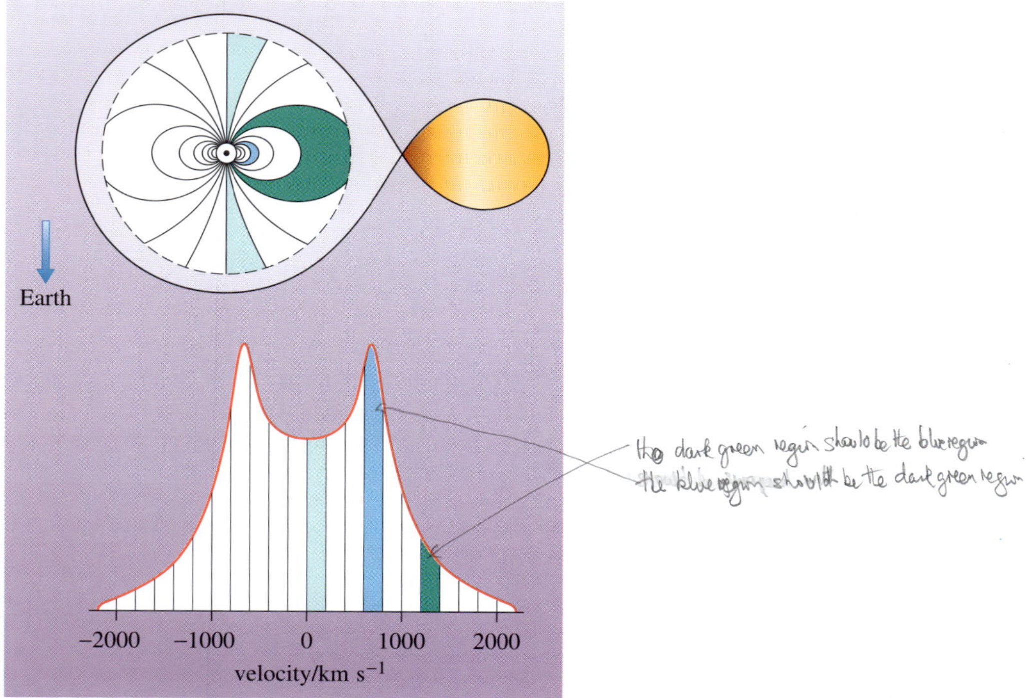

Figure 53 (*Upper panel*) A schematic view of the accretion disc residing inside the accretor's Roche lobe at binary phase 90° (see Figure 54 overleaf for a definition of binary phase). The arrow shows the direction to the observer. There is a pattern of curves superimposed onto the disc. The curves connect points in the disc with the same line-of-sight velocity of the disc plasma (as seen from Earth). Disc plasma within a section of the disc that is bounded by two such lines (such as the shaded areas) has a line-of-sight velocity in between the two velocity values the two curves correspond to. (*Lower panel*) The profile of a Doppler-broadened spectral line emitted by the disc shown in the upper panel. The spectral line is shown as a function of 'velocity'. This can be translated into the more familiar wavelength scale λ using the equation for the Doppler effect: $v = c(\lambda - \lambda_0)/\lambda_0$ (Equation 44 in Block 1). Here λ_0 is the rest wavelength which corresponds to zero velocity. Emission in the different velocity bins arises from the regions in the disc that are marked in the same colour.

Question 54

What is the chief cause of the double-peaked structure of accretion disc emission lines? ■

Question 55

Calculate the maximum Doppler shift of the Hα line (relative to its laboratory wavelength of 656 nm) emitted from the outer edge of an accretion disc with radius $0.5R_\odot$ around a white dwarf with mass $0.8M_\odot$. What is the maximum Doppler shift at the inner edge of the disc (at a radius of 7×10^6 m)? Consider only the Keplerian motion of the plasma in the disc, i.e. neglect the binary motion. ■

Accretion disc emission lines are Doppler-broadened. Each wavelength in a broad spectral line belongs to a certain radial velocity. As the disc plasma orbits the central accretor on Keplerian orbits it is easy to determine the disc region which has this line-of-sight velocity with respect to the observer. Figure 53 illustrates the regions of constant line-of-sight velocity in the disc when seen at binary phase 90° (see Figure 54 overleaf for a definition of binary phase, or orbital phase).

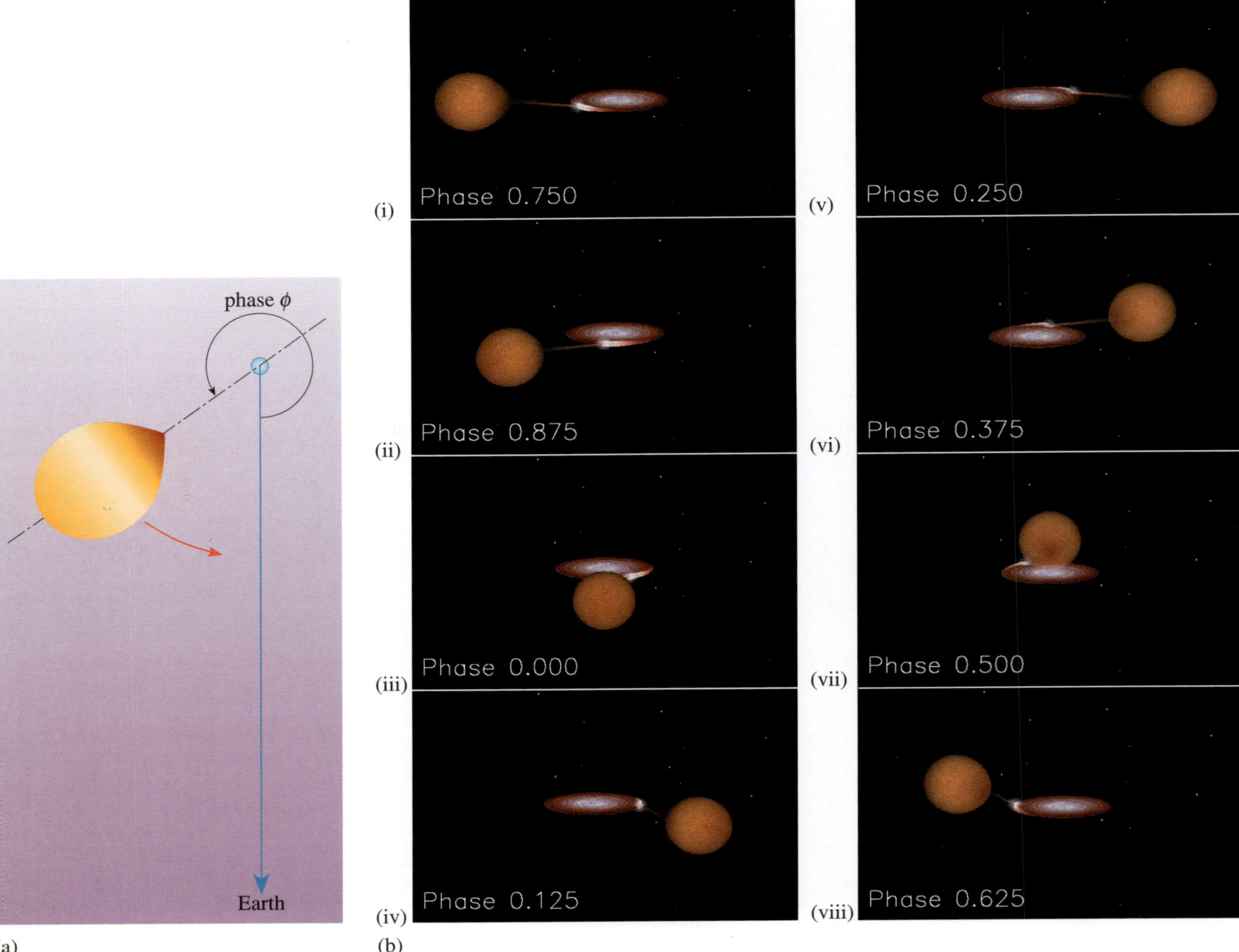

Figure 54 (a) The binary phase is the angle, as seen from the primary, between the instantaneous direction of the secondary star and the direction to the secondary star when it is closest to the observer. (b) The binary as seen by the observer for different binary phases, at an inclination of $i = 77°$. Phase 0 corresponds to $\phi = 0°$, phase 0.5 to $\phi = 180°$.

Question 56

Explain qualitatively why the pattern on the disc in Figure 53 is mirror-symmetric with respect to the x- and y-axis. ■

It is possible to learn a great deal about the nature of a binary star by analysing its **light curve**, the apparent variability of the emitted radiation as it completes one orbital revolution. The different viewing angle of the various luminous components in the system as a function of orbital phase (see Figure 54) causes the time variability of the observed flux.

The light curve can be measured in the integral (white) light, in certain colours, i.e. in restricted but still broad wavelength bands, or in narrow wavelength bands centred on certain spectral lines.

6.1 Eclipse mapping

The light curve of a binary star becomes particularly interesting when the system is eclipsing, i.e. when the more extended secondary star moves in front of the white dwarf and disc and blocks out its light.

■ What is the inclination angle of eclipsing systems?

❑ The line of sight must be close to the orbital plane, so the inclination angle is not much less than 90°. ■

Activity 32 **(15 minutes)**

Eclipsing CVs

Read the second and third paragraph on page 101 of FKR. From *The Energetic Universe* MM guide, watch the animation sequence 'Eclipsing CVs'. Look also at Figure 55.

Keywords: **bright spot** ■

FRANK KING RAINE

Activity 32

FKR p101

The Energetic Universe MM guide

'Eclipsing CVs'

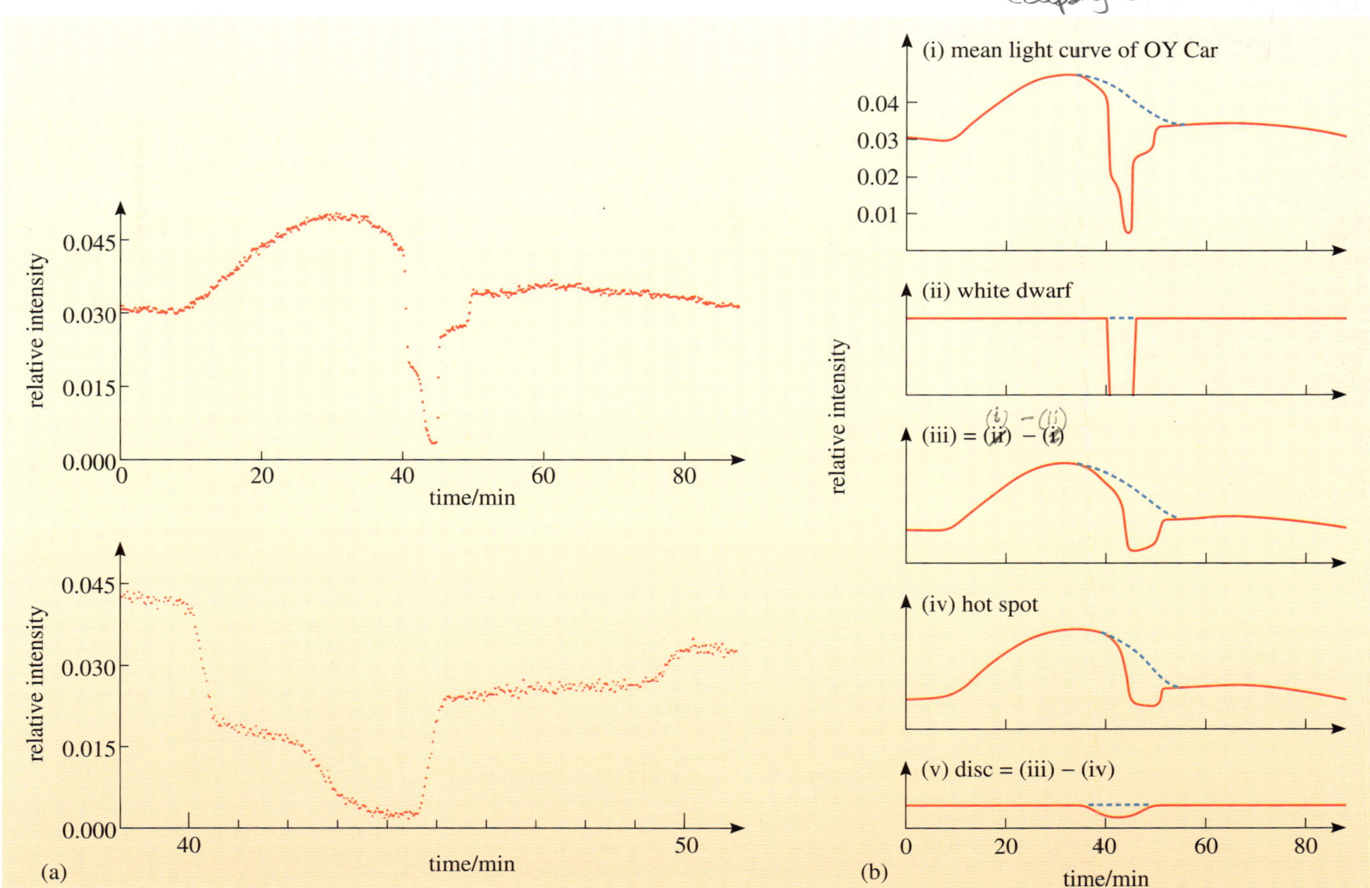

Figure 55 (a) Optical light curve of the eclipsing cataclysmic variable OY Car, showing a hot spot hump. (b) Decomposition of the eclipse light curve into its components: white dwarf, hot spot and accretion disc. The dotted curve shows how the light curve would appear in the absence of an eclipse.

Question 57

Explain what the bright spot (sometimes also called hot spot) is. At what orbital phase is the bright spot brightest? ■

Eclipsing systems offer the exciting opportunity of generating a map, almost a photograph, of the accretion disc as seen from above the orbital plane. How can this be achieved, given that all we see is a point source of light? The **eclipse mapping technique** exploits the fact that when the secondary's shadow moves across the disc it allows us to disentangle the contributions of different luminous parts of the disc from the integral light. Here 'shadow' denotes the disc regions which are hidden from our view because they are eclipsed by the secondary. By comparing the light from the CV when some fraction of it is blocked with the light from the uneclipsed disc we can deduce the contribution the eclipsed region made to the integral light. The shadow effectively probes different zones on the disc and therefore provides additional information needed to construct a two-dimensional map from a point source.

Activity 33

Eclipse mapping

Read the two paragraphs on page 103 of FKR (the second paragraph finishes on page 104).

Keywords: none ■

Figure 56 shows how the probing shadow moves across the disc in a system with inclination 78° and mass ratio 0.3.

Question 58

How does the length of the donor's shadow in the middle panel of Figure 56 depend on the inclination? ■

As you will no doubt have recognized, the eclipse mapping technique is a rather indirect method and requires a good deal of computer simulation to construct an image that is consistent with the observed eclipse light curve. In fact, one usually starts with a guessed model image and calculates how the light curve would look if the model were a perfect representation of the disc. The model light curve from this first guess is then compared to the actually observed light curve. If model and data do not agree the constructed model image is modified, and a new model light curve obtained. The correction procedure is repeated until the match between model and real light curve is satisfactory. The quality of fit in this iterative, step-by-step inversion method is determined by a statistical quantity that measures the overall deviation between data and model. Quite often this is the so-called χ^2 value ('chi-squared' value).

The Greek letter χ is pronounced 'kie', to rhyme with 'fly'.

The one-dimensional light curve does not in general uniquely constrain the two-dimensional brightness distribution in the disc. Hence the fitting procedure must impose an additional constraint, the closeness of the model to a default image. The default image represents an educated guess of the final image and should contain all essential features that one would expect to find, e.g. axis-symmetry except for a bright spot, and a radial brightness gradient. The fitting technique finds the model image that is most like the default image but still consistent with the data. The closeness of default and model image is measured by a numerical quantity called the image entropy (*not* the entropy of a gas, as used in thermodynamics!), and the fitting involves maximizing this quantity. Similar **maximum-entropy** imaging techniques are widely used in the reconstruction of images from incomplete data.

In the next activity you will take a closer look at real data and eclipse maps derived from them.

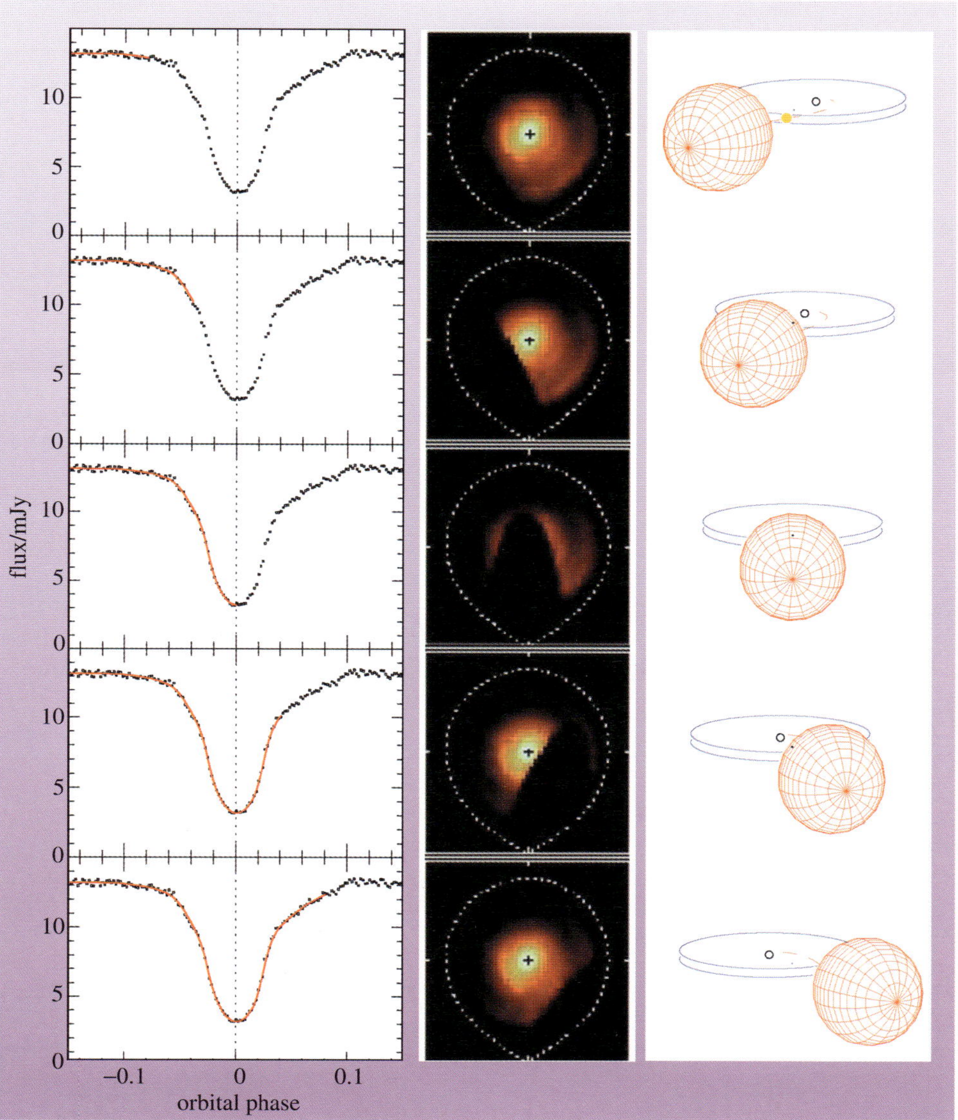

Figure 56 Simulated eclipse of an accretion disc, from binary phase −0.08 to binary phase +0.08. The right-hand panel shows how the binary star would appear from Earth if we had a powerful enough telescope to resolve it. The middle panel is a view from above the disc plane, with the Roche lobe of the accretor and the shadow of the secondary star indicated. The left panel shows the actually observed light curve (dots) and the model light curve (red line), reconstructed from the assumed brightness distribution in the accretion disc as shown in the middle panel. The red model light curve is only draw up to the respective binary phase. (Courtesy of Raymundo Baptista)

Activity 34 **(1 hour)**

The accretion disc in UU Aquarii

Connect to the S381 home page, select Block 3, Activity 34 and follow the instructions on screen on how to download the paper entitled 'Multicolour eclipse studies of UU Aquarii – II. The accretion disc' by R. Baptista, J. E. Steiner and K. Horne. The paper appeared in 1996 in the *Monthly Notices of the Royal Astronomical Society*, issue 282, pages 99–116.

Read the Introduction, Section 2, Section 3.1 and Section 4.3 up to the middle of the right-hand column on page 107. Do not bother about technical details. Focus on the bigger picture, and try to understand the content of Figures 1, 2, 3 and 8.

Keywords: none ■

Question 59

Explain the difference between the 10 light curves shown in Figure 2 of the paper on UU Aquarii. ■

Question 60

From radial temperature profiles such as those shown in Figure 8 of the paper it is possible to estimate the mass transfer rate in the binary. Describe how this is done. ■

6.2 Doppler tomography

A related and perhaps even more powerful imaging technique is **Doppler tomography**. The phase-dependent velocity information in emission lines replaces the shadow as a probe of different zones in the binary. As you have seen above in the context of double-peaked emission lines, the wavelength offset from the laboratory wavelength measures the radial velocity of the emitting material.

6.2.1 Velocity space

Consider the emission of a point-like source that is fixed in the co-rotating binary frame, e.g. a certain point on the surface of the Roche-lobe filling donor star. The radial velocity of this point would change sinusoidally with orbital phase, hence the corresponding emission line would trace out a sinusoid – an **S-wave** – when plotted against binary phase (Figure 57).

If there is more than one luminous spot in the system, the resulting phase-dependent spectral line is a superposition of individual S-waves, one for each point. The amplitude of these individual S-waves, and the binary phase at zero velocity, depend on the location of the corresponding spot in the system. Figure 58 shows how certain features of a semidetached binary as seen in the co-rotating binary frame correspond to features in velocity space, i.e. a coordinate system where the velocity in the y-direction is plotted against the velocity in x-direction. The following example should help you to understand Figure 58.

Example 20

(a) Explain how the point labelled '1' in Figure 58a relates to the point labelled '1' in Figure 58b.

(b) What are the position coordinates of the point with velocity coordinates $(v_x, v_y) = (0, 0)$?

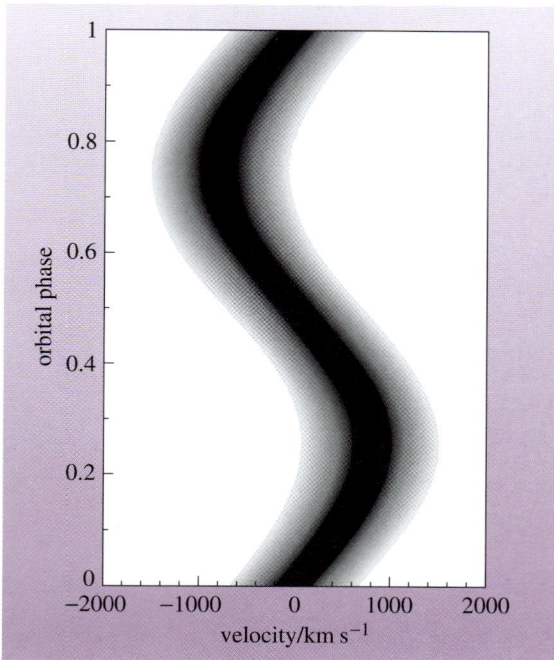

Figure 57 The S-wave from a luminous spot fixed in the binary frame is obtained by plotting the spectrum horizontally (wavelength is expressed as velocity, intensity indicated as brightness; the image is a 'negative', i.e. the darkest regions indicate the highest intensity), as a function of orbital phase in the vertical direction. The centre of the spectral line traces out a sinusoid.

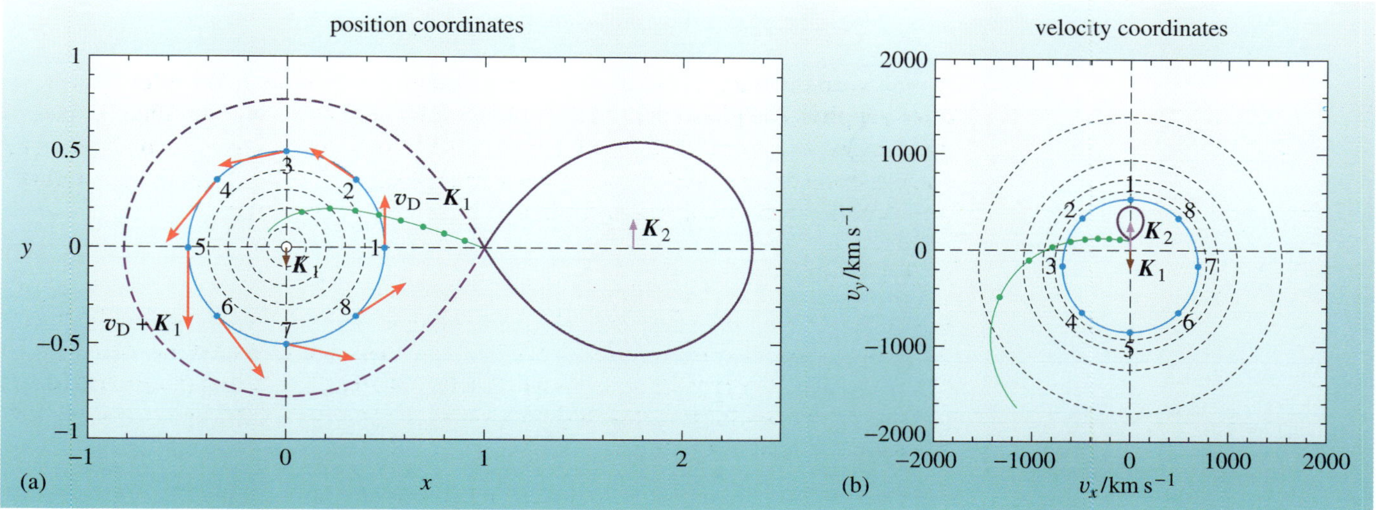

Figure 58 Position coordinates and velocity coordinates in a semidetached binary with accretion disc. Example 20 provides a detailed explanation of this figure.

Solution

Figure 58a shows the orbital plane (the x–y plane) of an interacting binary with the familiar Roche lobes. The circles around the accretor (in the left lobe; star 1) indicate the orbital motion of plasma in a Keplerian accretion disc around star 1. The red arrows along the circle with various points labelled by numbers indicate the velocity of the plasma flow, as seen from Earth. The velocity at each of the labelled points is the sum of two vectors,

$$\boldsymbol{v} = \boldsymbol{v}_\mathrm{D} + \boldsymbol{K}_1$$

where $\boldsymbol{v}_\mathrm{D}$ is the velocity of the plasma on its Keplerian orbit, i.e. with magnitude

$$|\boldsymbol{v}_\mathrm{D}| = v_\mathrm{K} = \sqrt{\frac{GM_1}{R}}$$

(where M_1 is the mass of the accretor and R the radius of the orbit) and direction tangential to the circle. The vector $\boldsymbol{K}_1$ is the orbital velocity of star 1 and, for the binary phase of 90° shown in this figure, points straight at the observer.

Figure 58b shows the same orbital plane, but in velocity coordinates. Consider a certain point in the orbital plane with coordinates (x, y) in Figure 58a. Suppose the stellar plasma at this point moves with velocity $\boldsymbol{v} = (v_x, v_y)$, where v_x is the x-component and v_y the y-component of $\boldsymbol{v}$. Then this point will appear as a point in Figure 58b at a position given by these velocity components, i.e. at (v_x, v_y).

At point 1, the velocity $\boldsymbol{v}_{\mathrm{D}}$ is in positive y-direction, i.e. opposite to the direction of $\boldsymbol{K}_1$. Accordingly, the vector sum $\boldsymbol{v} = \boldsymbol{v}_{\mathrm{D}} + \boldsymbol{K}_1$ has a vanishing x-component ($v_x = 0$), while the y-component is just the difference of the magnitude of $\boldsymbol{v}_{\mathrm{D}}$ and of $\boldsymbol{K}_1$, i.e. $v_y = v_{\mathrm{D}} - K_1$. Therefore, point 1 re-appears in Figure 58b at the point with coordinates $(v_x, v_y) = (0, v_{\mathrm{D}} - K_1)$. As $v_{\mathrm{D}} \gg K_1$ we have $v_y > 0$.

(b) The only point which does not move with respect to the observer is the centre of mass. This point would appear at $(v_x, v_y) = (0, 0)$. (This assumes that the binary system as a whole does not have a velocity relative to the observer.) ■

■ Where does the accretor appear on the velocity map?

□ The white dwarf moves in negative y-direction with speed K_1. So its velocity coordinates are $(v_x, v_y) = (0, -K_1)$. The white dwarf appears in the centre of the circle that connect the points labelled 1 to 8. ■

Do not worry if you still feel a bit uneasy about the translation between velocity space and real space – we will come back to this problem in Activity 36 and Question 62.

In practice, the luminous surfaces in a binary are not point-like but extended. But they can be thought of as a superposition of a large number of luminous spots. The Doppler tomography technique reconstructs the brightness distribution over the surfaces these points make up from the fine-structure in the observed broad S-wave. This imaging technique is also an inversion method, and maximum-entropy methods are applied abundantly. Usually one constructs an image in velocity space first. The translation into real space follows only once the model image satisfactorily fits the data in velocity space. It is sometimes easier to discuss the results in velocity space altogether, as this is independent of uncertainties and imperfections of the translation.

In Activity 35 you will see how the surface of a Roche-lobe filling star can be mapped. The technique is called Roche tomography.

6.2.2 Mapping the donor star and the disc

Activity 35 **(30 minutes)**

Activity 35
The Energetic Universe MM
Roche tomography

Roche tomography

From *The Energetic Universe* MM guide, watch the animation sequence ‘Roche tomography’. The image of the stellar surface generated by the Roche tomography technique reproduces the real surface features (dark starspots in a familiar shape) rather well.

Keywords: **Roche tomography**, **starspots** ■

Question 61

Is the amplitude of the S-curve of a starspot on the equator of the secondary bigger or smaller than the amplitude for a starspot on the pole of the secondary? Explain why. ■

MEDICAL IMAGING: THE CT SCANNER

The imaging techniques discussed in this section – eclipse mapping, Roche tomography and Doppler tomography – are reminiscent of a medical imaging technique you might be familiar with: the computed tomography (CT) scanner. A CT scan makes use of the attenuation of X-rays that pass through the patient: different parts of the body are more or less transparent to X-rays, causing a pattern of varying shades on the X-ray image of the body. A CT scan involves the imaging of a succession of two-dimensional slices through the body (*tomos* is Greek for 'slice') to build up three-dimensional information of the body (see Figure 59).

Each two-dimensional slice of the CT scan is exposed to a probing X-ray beam that rotates around the body. This way a 'projection' of the slice, i.e. an edge-on view of the slice in X-rays, can be obtained from different angles. By combining the information contained in the different projections, the full two-dimensional image of the slice can be reconstructed by a computer.

This is in close analogy to the eclipse mapping technique where the probing shadow of the secondary moves across the disc due to the binary motion, and the binary light is registered as a function of orbital phase. The two-dimensional disc structure is reconstructed from the information contained in the light curve.

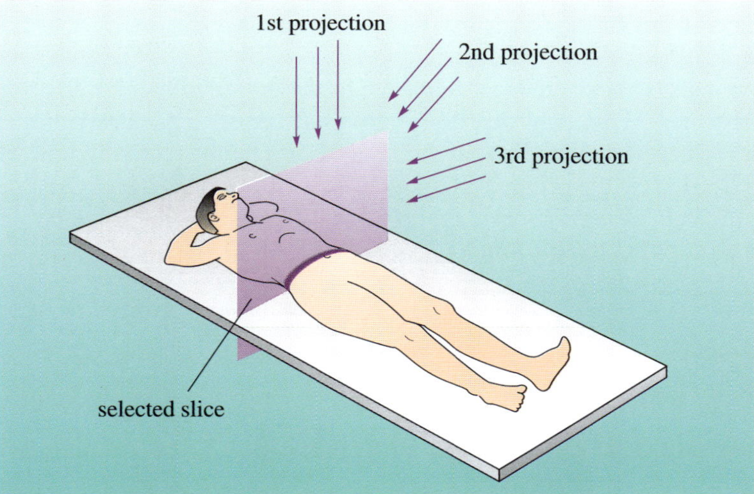

Figure 59 Principle of a CT scanner in medical imaging. X-ray beams are passed through a slice of the body in different directions.

The situation becomes a bit more complicated when the emitting gas is moving in the binary frame. Of course, this is just what happens for accretion disc material on Keplerian orbits around the primary star. It is still possible to construct an image in velocity space, as before, with all the complications and problems attached to it. The translation into real space is even more complicated as it introduces further ambiguity. The same velocity pattern can arise from emitting gas at very different locations in the binary, with very different velocities with respect to the binary frame. If the construction of the velocity map is complicated and involved, the interpretation of the map can be an art!

Activity 36 **(1 hour)**

Doppler tomography

Connect to the S381 home page, select Block 3, Activity 36 and follow the instructions on screen on how to download the paper *Images of accretion discs – II. Doppler tomography*, by T. R. Marsh and K. Horne. This appeared in *Monthly Notices of the Royal Astronomical Society*, 1988, Vol. 235, pages 269–286. Read Sections 1 and 2.

This is the pioneering paper which introduced the Doppler tomography technique as a powerful tool in accretion disc research. As such it is written for researchers, not for students. Try to extract the basic principles, and try to understand Figures 1 and 2.

Keywords: **velocity coordinates**, **position coordinates** ■

Question 62

The velocity map of an accretion disc turns the disc 'inside-out'. Explain this statement. ■

A remarkable success story of the Doppler imaging technique came in 1996 with the discovery of a spiral shock pattern in the accretion disc of IP Pegasi (Figure 60 and Figure 61). These spiral shocks are a consequence of the gravitational perturbations from the secondary star on the accretion disc. In our theoretical considerations above we neglected any effects the secondary might have other than the restriction on the size of the disc imposed by the primary's Roche lobe. Spiral shocks have been predicted by numerical calculations of accretion discs. The shocks are roughly stationary in the co-rotating binary frame.

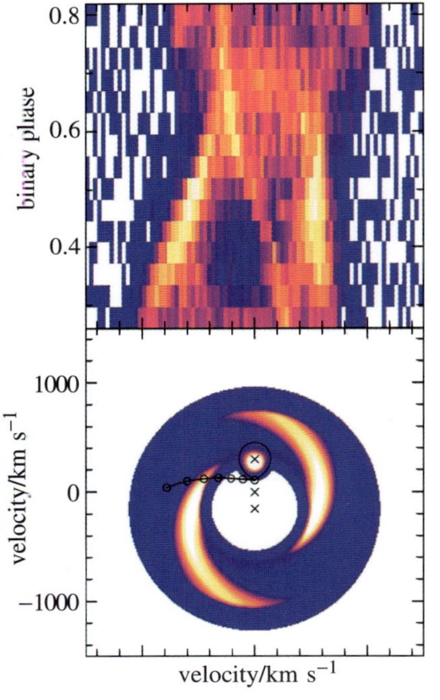

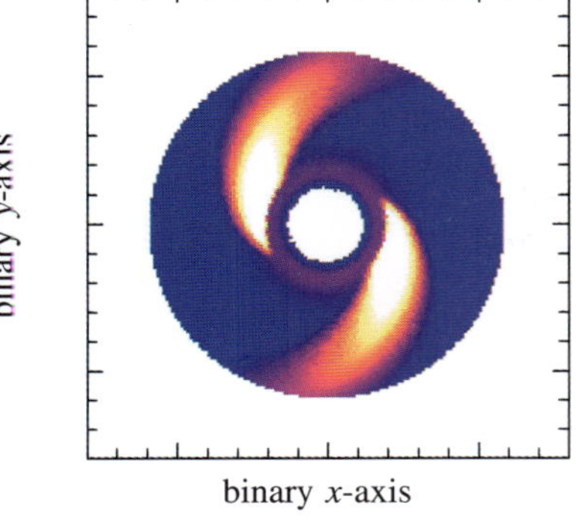

Figure 60 A synthetic (model) spiral shock pattern in an accretion disc. (*Lower panel*) In the binary frame (position coordinates). (*Middle panel*) In velocity space. (*Upper panel*) The corresponding S-wave pattern, the so-called trailed spectrum. The crosses in the middle panel denote the centre of the donor star, the centre of mass, and the centre of the white dwarf (from top to bottom). The open circles, connected by a curve, show the location of the mass transfer stream.

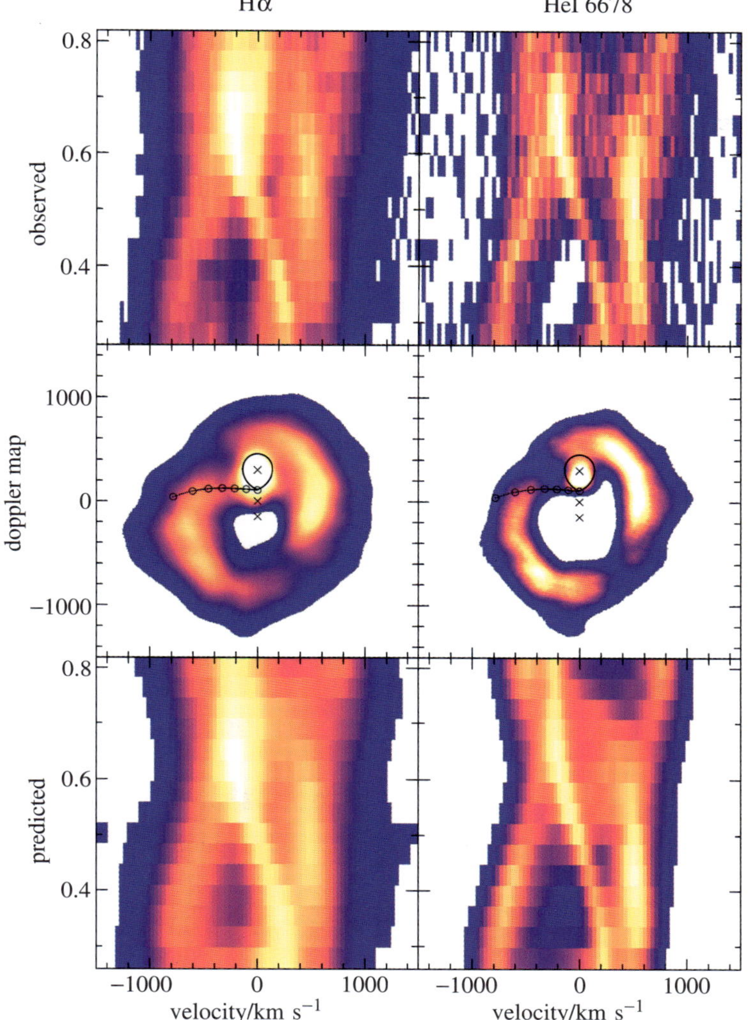

Figure 61 A Doppler tomogram of spiral shocks in the cataclysmic variable IP Pegasi. (*Left*) Hα emission line; (*Right*) HeI line at 667.8 nm. The upper panel shows the actually observed trailed spectra (S-wave pattern), the middle panel shows reconstructed images in velocity space. The 'predicted' trailed spectra in the lower panel are obtained from this best fit and serve to check the method.

6.3 X-ray discs

The observational evidence from CVs largely verifies the basic principles of accretion disc physics that we have considered in the earlier sections. Clearly we expect that the same principles hold for the neutron star and black hole cousins of CVs, the low-mass X-ray binaries. However, the situation here is more complicated and many of the clear-cut disc signatures we met in CVs are absent from X-ray binaries.

Activity 37 **(20 minutes)**

Discs in X-ray binaries

Read from FKR page 104, starting at 'For systems other than cataclysmic variables …', up to and including the first paragraph on page 108. Then jump to page 110 and read the last paragraph of Section 5.7.

Keywords: **X-ray corona, X-ray irradiation, X-ray dips, warped discs** ■

Question 63

The lack of eclipses in bright X-ray binaries has been attributed to a selection effect. Repeat the argument in your own words. ■

Question 64

Explain what X-ray dips are and how they occur. What is the difference between X-ray dips and partial X-ray eclipses by the secondary star? ■

Question 65

X-ray dips are observed in both eclipsing and non-eclipsing X-ray binaries. What does that say about the size of the structure producing the dips with respect to the secondary star? ■

The relative lack of eclipsing X-ray binaries, and the occurrence of X-ray dips in non-eclipsing systems, all point towards the existence of disc material high above the orbital plane (see also Figure 62), perhaps as high as $H \approx 0.2 \times R$ if R is the distance from the neutron star or black hole.

- ■ How thick are typical CV accretion discs?
- ❑ The relative disc scale height H/R is of order of the local sound speed divided by the Keplerian speed, i.e. typically less than a few per cent (see also Equation 67). This is more than a factor of 10 less than in X-ray binaries. ■

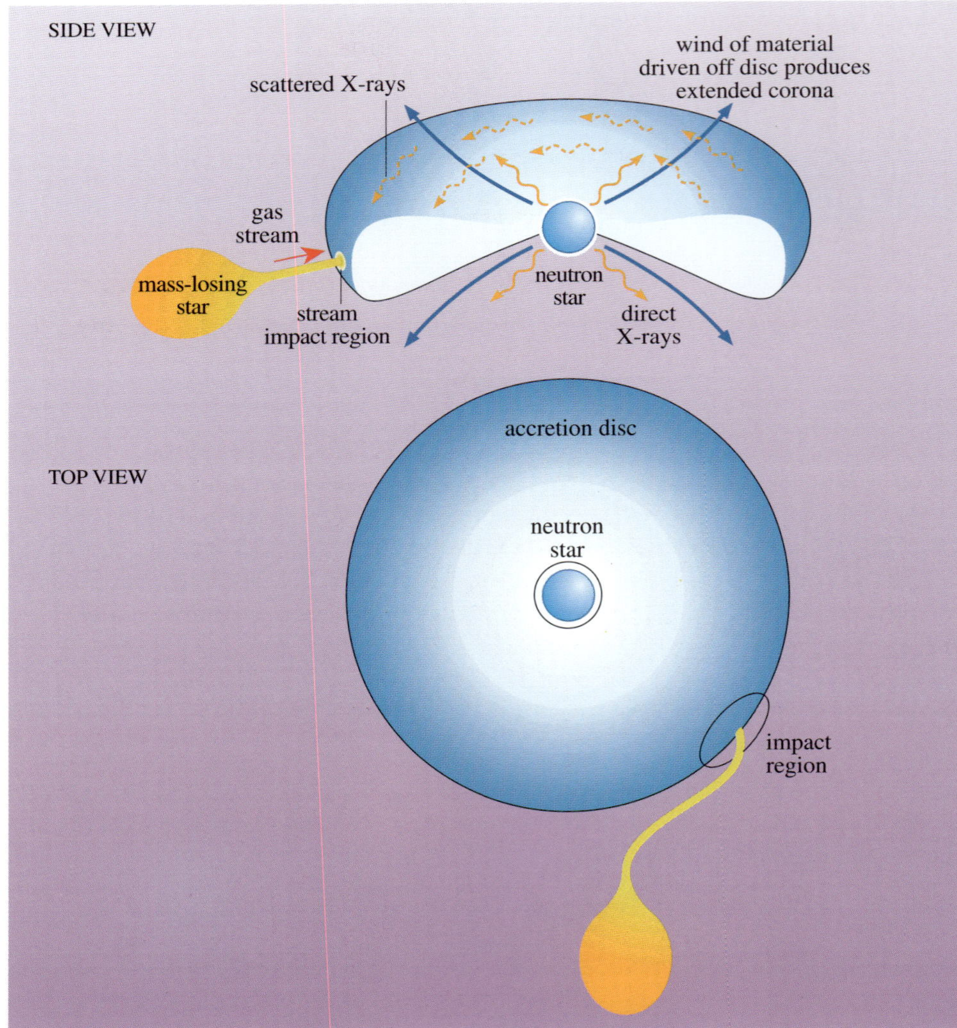

Figure 62 Schematic view of the disc structure and X-ray emission in an X-ray bright low-mass X-ray binary.

So it is almost as if the discs in X-ray binaries are much thicker than in CVs. One possible reason for why the discs may *appear* much thicker is that they could be warped. A **warped accretion disc** is, of course, no longer globally flat, and hence, strictly speaking, no longer a 'disc'. But *locally* it is still flat. You can see what we mean by a warped disc from the following recipe of how to make one. Break the initially flat disc up into many narrow disc rings. Then tilt each ring by a small angle δ against the orbital plane, and rotate the tilted ring about the rotation axis of the original disc by an angle ε (Figure 63). If ε is the same for each ring the resulting structure is of course again a flat disc, inclined against the orbital plane by the angle δ. But if ε is a function of the ring radius, e.g. linearly increasing with R, the result is a tilted and twisted disc.

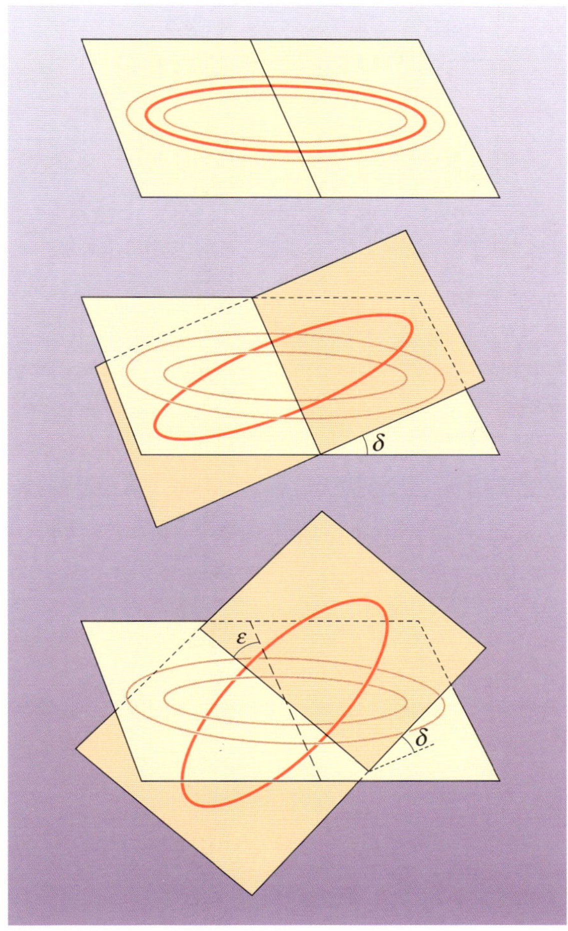

Figure 63 Constructing a warped disc: the angles δ and ε.

If the angle δ is large enough, the disc might appear rather thick to a distant observer who cannot resolve the detailed, warped structure – even though the initial disc was perfectly thin (Figure 64 overleaf). A physical mechanism which can cause the disc to warp is an instability that arises when the disc is irradiated. The irradiating luminosity will be re-radiated or scattered from the disc surface. The resulting radiation pressure causes a reaction force. If there is a deviation from complete symmetry about the orbital plane a net torque results, and a warp will form.

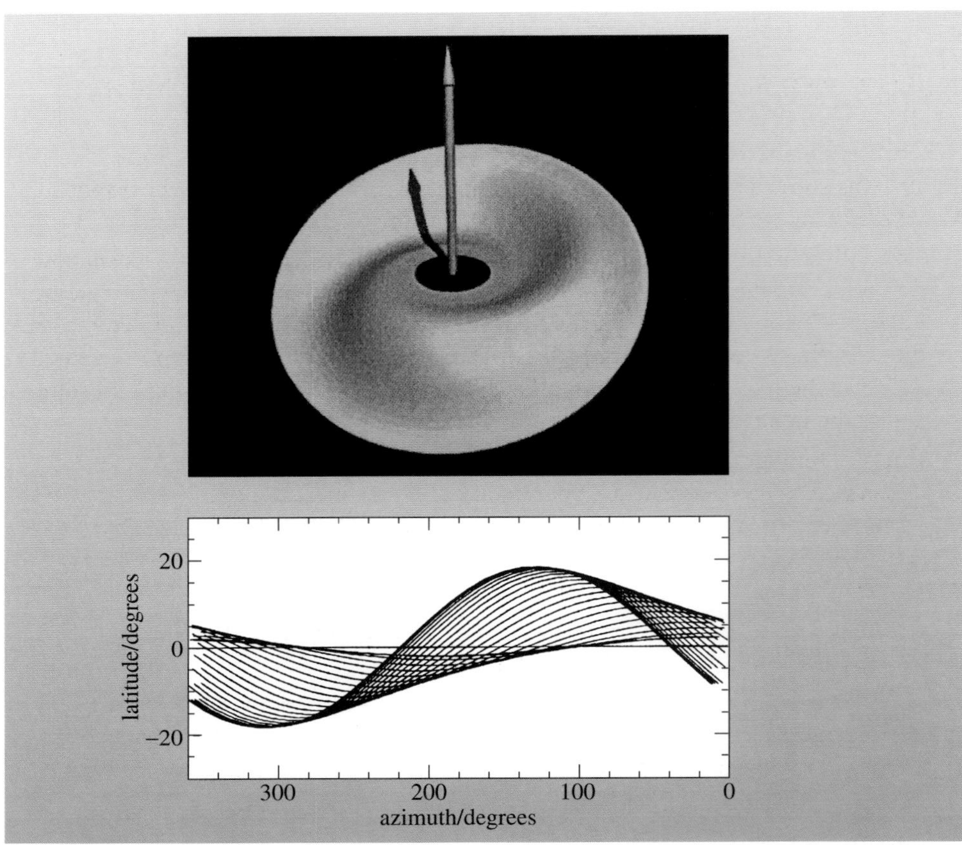

Figure 64 Shape of a warped disc (top), and an edge-on view of the disc (bottom).

Question 66

In the above recipe for tilted and twisted discs, how large a tilt angle δ is necessary to explain an apparent disc of thickness $H/R = 0.2$, if the underlying disc is infinitesimally thin? ■

We have now pretty much exploited the theory of steady-state accretion. To appreciate the full richness of energetic phenomena that occur in accreting binary stars we obviously have to go a step further. You know already that many accreting binary systems exhibit truly dramatic changes of the observed flux with time – just think of dwarf novae or soft X-ray transients (Section 1.3). Clearly, to understand these we have to consider time-dependent accretion. The physics of time-dependent accretion can be hugely more complicated than the physics of steady-state accretion. Don't panic – the path we shall take to approach this will make abundant use of what we learned about steady-state discs, and the physics we derive in this way will represent the climax of this block. But before we embark on this exciting undertaking you can refresh your brain cells on a relaxing detour into the world of magnetic stars, where accretion can take on a very different form.

The name game: part 6

Skim-read Section 6 and make a list of all symbols that denote *new* variables and *new* quantities, and note down what they mean. Flag up those symbols that have (unfortunately) multiple uses.

Once you have compiled your list and explained the meaning of all entries, compare it to the one given in Appendix A5 at the end of the Study Guide.

6.4 Summary of Section 6

1. A clear observational signature of accretion discs is double-peaked emission lines. They reflect the Keplerian motion of the plasma in the disc.
2. The light curve of accreting binaries is a powerful diagnostic tool. Some systems show a pronounced orbital hump, caused by the bright (or hot) spot where the mass transfer stream from the secondary impacts the disc.
3. The eclipse mapping technique exploits the fact that as the secondary's shadow moves across the disc it covers and uncovers the different luminous parts in the system. The one-dimensional light curve does in general not uniquely constrain the two-dimensional brightness distribution in the disc. The fitting procedure enforces the likeness of the model to a default image, by maximizing a quantity called the image entropy.
4. Doppler tomography is a powerful imaging technique that uses the phase-dependent radial velocity information in emission lines to probe different zones in the binary. Using maximum-entropy techniques the brightness distribution in the binary is reconstructed from the fine structure in the observed S-wave of spectral lines. The reconstruction gives an image in velocity space.
5. Roche tomography is the application of Doppler tomography to map the surface structure of the Roche-lobe filling star.
6. A recent success of Doppler tomography has been the discovery of a spiral shock pattern in some accretion discs. The spiral shocks are a consequence of gravitational perturbations of the secondary star on the accretion disc.
7. The light curves of X-ray binaries point to the existence of an extended X-ray emitting corona above the disc.
8. The relative lack of eclipsing X-ray binaries, and the occurrence of X-ray dips even in non-eclipsing systems show that there is disc material high above the disc mid-plane. Discs in X-ray binaries may be warped, because of an instability when the disc is irradiated.

7 MAGNETIC STARS: ATTRACTIVE OR REPULSIVE?

Magnetic fields are very common in astrophysical systems. You are certainly familiar with the effect of the Earth's magnetic field that causes a compass to point to the magnetic North Pole. Such large-scale fields are particularly spectacular when they interact with a plasma flow. Cosmic plasma essentially consists of ionized hydrogen, a mixture of protons and electrons. These charged particles feel a magnetic force – the Lorentz force (see Equation 139 in Block 1) – which will in general deflect the particles from their original path. The magnetic force is perpendicular to both the instantaneous motion of the ions and the direction of the local magnetic field. As a consequence individual ions follow tightly wound helical paths around field lines (see, for example, Figure 71 below). The macroscopic effect of this is that plasma can move essentially freely along field lines, but not perpendicular to them. If an external effect enforces the bulk motion of plasma perpendicular to the field lines (e.g. when the accretion stream crashes into the field of a magnetic white dwarf) then the moving medium avoids crossing the field lines by dragging them along with it. A moving, charged medium can transport magnetic field lines!

This is the basis of an enormously useful concept in **magneto-hydrodynamics**, the study of the flow of an electrically conducting fluid (i.e. a plasma) in the presence of a magnetic field:

> Magnetic fields are 'frozen' in to a perfectly conducting plasma.

The plasma and the field move exactly together, in the sense that a magnetic field line (see the box on 'Magnetic field lines') is traced out by the same plasma particles at all times. This concept is not just a helpful illustration of phenomena in magneto-hydrodynamics, it is indeed quite often used as a mathematical tool to quantitatively describe these phenomena. The concept of frozen field lines can be derived in a rigorous way from the fundamental laws that apply in the classical discipline of electromagnetism, the Maxwell equations, but this is beyond the scope of this course.

7.1 The magnetospheric radius

Our Sun offers many beautiful examples that illustrate frozen field lines. Figure 65 shows an arc-shaped structure of hot plasma that extends above the solar photosphere. It is supported by a similarly arced magnetic field. The intricate fine structure of such plasma loops provides a glimpse of the awesome complexity of magnetic field patterns on the surface of the Sun, such as the one shown in Figure 66.

Figure 65 A large solar prominence as seen by the SOHO satellite at 30.4 nm wavelength, probing the emission of ionized helium. A prominence is a dense, relatively cool cloud of plasma suspended in the hot, thin solar corona.

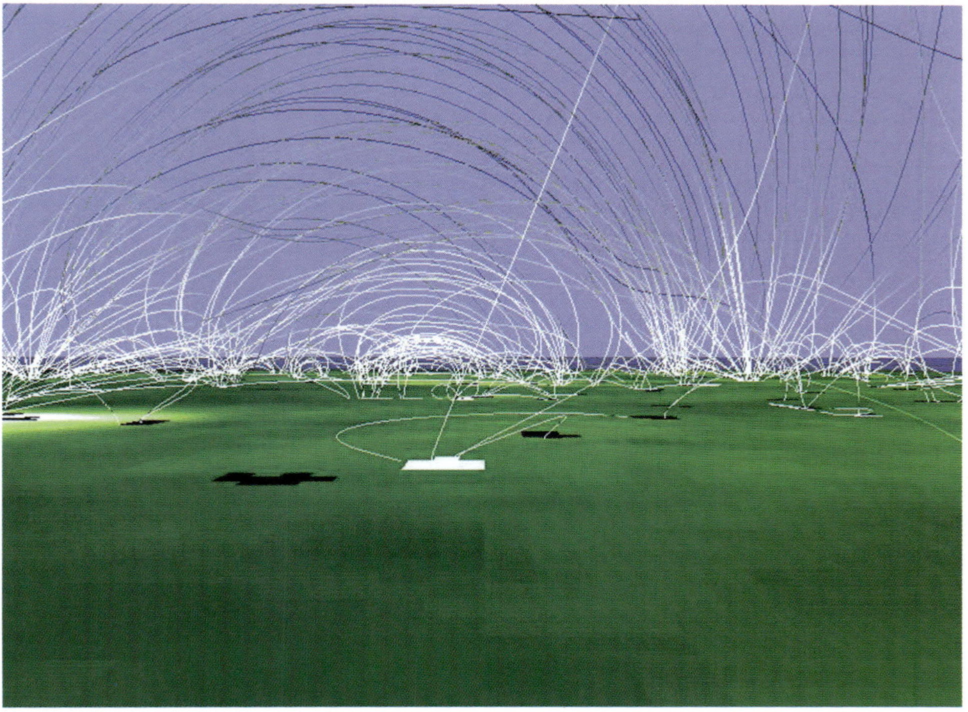

Figure 66 Reconstructed magnetic field structure ('magnetic carpet') on the surface of the Sun.

MAGNETIC FIELD LINES

The significance of a magnetic field line is that at each point in space on the line the magnetic flux density vector $\boldsymbol{B}$ is tangential to the field line. (We shall often use the terms *magnetic field* or *magnetic field strength* when we refer to $\boldsymbol{B}$. We note that the proper designation for $\boldsymbol{B}$ is *magnetic flux density*. Fortunately the difference between flux density and field strength is not important for our applications.) The density of field lines in a given volume of space is indicative of the magnitude of $\boldsymbol{B}$ at this point in space. The magnetic pressure is large when the density of field lines is high.

A convenient way to quantify the relative importance of a magnetic field for the fluid flow is the concept of **magnetic pressure**. Plasma in a magnetic field $\boldsymbol{B}$ with magnitude B behaves as if it is subject to an additional pressure, given by (in SI units)

$$P_{\text{mag}} = \frac{4\pi}{\mu_0}\frac{B^2}{8\pi} \tag{70}$$

This has to be compared to the gas pressure from the ideal gas law:

$$P_{\text{gas}} = \rho\frac{k}{m_{\text{H}}}T = \rho c_{\text{s}}^2$$

and in particular to the inertia of the gas moving with bulk velocity v, the **ram pressure**, $P_{\text{ram}} = \rho v^2$.

■ Verify that ρv^2 has the dimension of a pressure.

❑ The SI unit of ρ is kg m^{-3}, the unit of v is m s^{-1}, hence ρv^2 has the unit $\text{kg m}^{-3}\,\text{m}^2\,\text{s}^{-2} = \text{kg m}^{-1}\,\text{s}^{-2}$. This is also the unit of pressure, as force per area has the unit $\text{N m}^{-2} = \text{kg m s}^{-2}\,\text{m}^{-2} = \text{kg m}^{-1}\,\text{s}^{-2}$. ■

The gas pressure arises from the random motion of ions (or atoms, molecules) within the gas, while the ram pressure is due to the bulk motion of the gas. The ram pressure is the force exerted on a unit surface area facing the gas stream, resulting from the linear momentum carried in the flow.

These different types of pressure measure the relative importance for the plasma flow of the three different effects that are causing them. A high magnetic pressure signals a strong magnetic field. The gas pressure is high if the gas is hot and dense. A large ram pressure implies a high bulk speed. Hence if P_{mag} is the dominant pressure, then the magnetic field determines the fluid velocity. The plasma follows the fixed field lines. If P_{ram} dominates, the magnetic field is effectively dragged around by the fluid, and the fluid essentially behaves as if the magnetic field were absent. In the intermediate case, $P_{\text{gas}} \gg P_{\text{mag}} \gg P_{\text{ram}}$, the magnetic field moves the fluid, but is itself confined by gas pressure.

Accretion onto compact stars is particularly liable to magnetic effects, because many compact objects have a large surface field strength. It is thought that the magnetic flux through the surface of a star ($\propto B_* R_*^2$) is roughly conserved throughout the star's life. So a normal star with radius R_* of order a solar radius and a rather modest surface field strength B_* is likely to end up as a highly magnetic (large B_*) compact (small R_*) star.

Question 67

(a) Our Sun will evolve into a $0.6M_{\odot}$ white dwarf in about 5×10^9 years. What will be the surface magnetic field strength of this white dwarf star? Assume that the Sun's global average surface field is 1 gauss, and that the white dwarf radius is 8.7×10^8 cm. (b) How magnetic are red giants? ■

Gauss (G) is a cgs unit, the SI unit is tesla (T). 1 T = 10^4 G.

The simplest and most common global magnetic field pattern is that of a **magnetic dipole**, similar to the field of a small bar magnet (Figure 67).

From the field line density shown in Figure 67 it is clear that the field strength is highest on the surface of the star, and reduces when one moves away from it. The field strength varies roughly as

$$B \approx \frac{\mu}{r^3} \tag{71}$$

with distance r from the centre of the star. The constant μ is called the magnetic dipole moment, and can be calculated from $\mu = B_* R_*^3$ where B_* is the (average) surface magnetic field strength, and R_* the radius of the star. Note that Equation 71 describes the field structure *outside* of a star with a dipolar field that is characterized by a fixed magnetic moment μ. This is *different* from the above statement about the conservation of magnetic flux $\propto B_* R_*^2$ on the surface of a star throughout the star's life, during which the magnetic moment may *vary*.

Be careful not to confuse the magnetic moment μ with the permeability as in Equation 70!

■ Why does magnetic pressure drop with the sixth power of the distance from the accreting star?

❑ The accretor has a dipolar field $B \propto r^{-3}$. Magnetic pressure is proportional to B^2, hence proportional to r^{-6}. ■

(a)

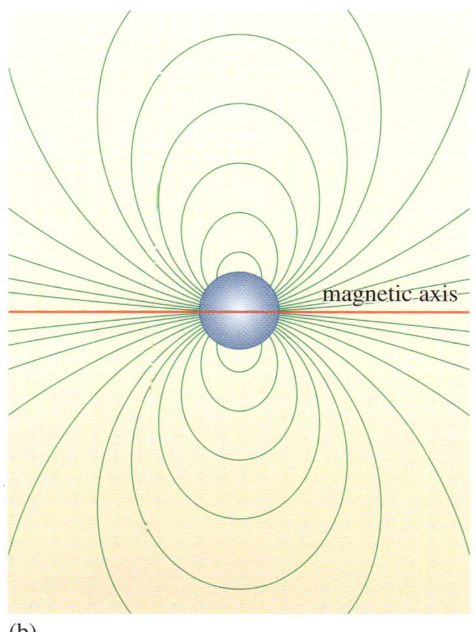

(b)

Figure 67 (a) Magnetic field lines of a magnetic dipole, and (b) a star with a dipolar magnetic field.

Equation 71 is only an approximation as the field line pattern is obviously not spherically symmetric. In particular, for a given μ, the field strength on the surface varies with position, and is highest close to the magnetic poles.

Question 68

Many pulsars (neutron stars) have a surface magnetic field of order 10^{12} G. Given that a typical neutron star radius is 10 km, calculate the corresponding magnetic moment and compare it to the typical magnetic moment $10^{34}\,\mathrm{G\,cm^3}$ of magnetic white dwarfs in polars. ■

You are now well prepared to investigate the accretion onto a magnetic compact star in some detail. In the following activities you will read selected paragraphs of Section 6.3 in FKR.

Activity 38 (20 minutes)

Alfvén radius

Read Section 6.3 of FKR up to the middle of the paragraph on page 160 following FKR Equation 6.21; the last sentence is 'Thus we expect accretion on to neutron stars to be controlled by the magnetic field near the surface in many cases.'

Keywords: **Alfvén radius** ■

The magnitude of the free-fall velocity

$$v_{\mathrm{ff}} = \sqrt{\frac{2GM}{r}} \tag{72}$$

is the speed of a particle with a very small mass in free-fall towards a body with mass M, assuming that the particle has started from infinity at rest. The quantity r is the distance between M and the particle. The free-fall velocity is analogous to the Keplerian velocity (Equation 1).

■ So what is the difference between the free-fall velocity and Kepler velocity if $r = R$?

□ A factor $\sqrt{2}$ in magnitude and the direction. A free-falling particle is on a head-on collision course with the central body. The Kepler velocity refers to a Kepler orbit, i.e. in this case a circular orbit around the central body. So although the two velocities have almost the same magnitude, they are perpendicular to each other. ■

Question 69

Explain in words what the Alfvén radius is. ■

We retrace the steps FKR takes to derive an expression for the Alfvén radius in the following two examples.

Example 21

In order to derive the Alfvén radius, FKR referred back to FKR Equation 2.23,

$$4\pi r^2 \rho(-v) = \dot{M} \tag{73}$$

This is analogous to the mass conservation equation for accretion through a disc, but now for spherically symmetric accretion from all directions. Show how this equation is obtained.

Solution

We consider a spherical inflow of mass, with velocity $v < 0$ (by convention, the velocity is positive if the mass flows in the positive r-direction, i.e. away from the centre). The surface area of a sphere with radius r is $4\pi r^2$. In the time Δt all particles in a spherical shell with width $\Delta r = (-v)\Delta t$ just outside this sphere will cross the surface of the sphere. The volume of the shell is $\Delta V = 4\pi r^2 \Delta r$, hence the mass ΔM crossing the surface is $\Delta M = \rho \Delta V$. This corresponds to a mass accretion rate

$$\dot{M} = \frac{\Delta M}{\Delta t} = \frac{\rho \Delta V}{\Delta t} = \frac{\rho 4\pi r^2 \Delta r}{\Delta t}$$

We have $\Delta r/\Delta t = -v$ as Δr and $\Delta t \to 0$, hence

$$\dot{M} = \frac{\rho 4\pi r^2 \Delta r}{\Delta t} = 4\pi r^2 \rho(-v)$$ ■

Example 22

(a) Derive the expression FKR Equation 6.18 for the Alfvén radius, r_M.

(b) Verify that the expression is dimensionally correct.

Solution

(a) At the Alfvén radius the magnetic pressure equals the ram pressure, $P_\mathrm{mag} = P_\mathrm{ram}$.

Using Equation 71 to express the magnetic field strength in terms of the magnetic dipole moment, the magnetic pressure (Equation 70) can be written as

$$P_\mathrm{mag} = \frac{4\pi}{\mu_0}\frac{\mu^2}{8\pi r^6} = \frac{1}{\mu_0}\frac{\mu^2}{2r^6}$$

The ram pressure is $P_\mathrm{ram} = \rho v^2$, where v is the free-fall speed. We use Equation 73 and Equation 72 to rewrite this as

$$P_\mathrm{ram} = \rho|v| \times |v| = \frac{\dot{M}}{4\pi r^2} \times \left(\frac{2GM}{r}\right)^{1/2}$$

Hence we have, for $r = r_\mathrm{M}$,

$$\frac{1}{\mu_0}\frac{\mu^2}{2r^6} = \frac{\dot{M}}{4\pi r^2} \times \left(\frac{2GM}{r}\right)^{1/2}$$

This is $$\frac{1}{\mu_0}\frac{\mu^2}{2r^6} = \frac{(2GM)^{1/2}\dot{M}}{4\pi r^{5/2}}$$

hence $$\frac{1}{\mu_0}\frac{\mu^2}{2} \times \frac{4\pi}{(2GM)^{1/2}\dot{M}} = \frac{r^6}{r^{5/2}}$$

Rearranging the equation, this becomes

$$r^{7/2} = \left(\frac{2\pi}{\mu_0}\right)(2G)^{-1/2}\dot{M}^{-1}M_1^{-1/2}\mu^2 \qquad (74)$$

We now solve this for r:

$$r_\mathrm{M} = \left(\frac{2\pi}{\mu_0}\right)^{2/7}(2G)^{-1/7}\dot{M}^{-2/7}M_1^{-1/7}\mu^{4/7} \qquad (75)$$

To compare this with FKR Equation 6.18 we have to insert numbers (this looks huge, but it is no different from Equation 75):

$$r_M = \left(\frac{2\pi}{4\pi\,\text{dyne}\,\text{emu}^{-2}\,\text{s}^2}\right)^{2/7} \times (2\times 6.67\times 10^{-8}\,\text{dyne}\,\text{cm}^2\,\text{g}^{-2})^{-1/7}$$

$$\times\left(\frac{\dot{M}}{10^{16}\,\text{g}\,\text{s}^{-1}}\right)^{-2/7} \times (10^{16}\,\text{g}\,\text{s}^{-1})^{-2/7} \times \left(\frac{M_1}{1\text{M}_\odot}\right)^{-1/7}$$

$$\times (1.989\times 10^{33}\,\text{g})^{-1/7} \times \left(\frac{\mu}{10^{30}\,\text{G}\,\text{cm}^3}\right)^{4/7} \times (10^{30}\,\text{G}\,\text{cm}^3)^{4/7}$$

This is

$$r_M = \left(\frac{2\pi}{4\pi}\right)^{2/7} \times (2\times 6.67\times 10^{-8})^{-1/7} \times \dot{M}_{16}^{-2/7} \times (10^{16})^{-2/7}$$

$$\times\, m_1^{-1/7} \times (1.989\times 10^{33})^{-1/7} \times \mu_{30}^{4/7} \times (10^{30})^{4/7}\,\text{cm}$$

or finally $r_M = 5.1\times 10^8\,\dot{M}_{16}^{-2/7} m_1^{-1/7} \mu_{30}^{4/7}\,\text{cm}$

which is just FKR Equation 6.18.

(b) It is actually less tedious if we verify that the equation is dimensionally correct by checking Equation 74 (rather than doing it once both sides are taken to the power of 2/7 to solve for r). As $\mu_0 = 4\pi\,\text{dyne}\,\text{emu}^{-2}\,\text{s}^2$ the right-hand side of Equation 74 has the following units:

$$(\text{dyne}\,\text{emu}^{-2}\,\text{s}^2)^{-1} \times (\text{dyne}\,\text{cm}^2\,\text{g}^{-2})^{-1/2} \times (\text{g}\,\text{s}^{-1})^{-1} \times \text{g}^{-1/2} \times (\text{G}\,\text{cm}^3)^2$$

this is equal to

$$\text{dyne}^{-1-1/2}\,\text{emu}^2\,\text{s}^{-2+1}\,\text{cm}^{-1+6}\,\text{g}^{1-1-1/2}\,\text{G}^2 = \text{dyne}^{-3/2}\,\text{emu}^2\,\text{s}^{-1}\,\text{cm}^5\,\text{g}^{-1/2}\,\text{G}^2$$

Now dyne = $\text{g}\,\text{cm}\,\text{s}^{-2}$ and gauss = $\text{dyne}\,\text{emu}^{-1}\,\text{cm}^{-1}\,\text{s} = \text{g}\,\text{s}^{-1}\,\text{emu}^{-1}$, so that

$$(\text{g}\,\text{cm}\,\text{s}^{-2})^{-3/2}\,\text{emu}^2\,\text{s}^{-1}\,\text{cm}^5\,\text{g}^{-1/2}(\text{g}\,\text{s}^{-1}\,\text{emu}^{-1})^2 = \text{g}^{-3/2-1/2+2}\,\text{cm}^{-3/2+5}\,\text{s}^{+3-1-2}\,\text{emu}^{2-2} = \text{cm}^{7/2}$$

This is indeed also the unit of the left-hand side of the expression, $r^{7/2}$. ■

We have seen that magnetic CVs appear in two flavours. Firstly, there are the polars where the accretion stream couples immediately to the white dwarf's magnetic field once it leaves the secondary star. Obviously, in polars, the magnetospheric radius must be at least as large as the white dwarf's Roche lobe radius. On the other hand there are the intermediate polars (IPs), some of which are known to have an accretion disc. Hence in IPs the magnetospheric radius must be smaller than the Roche lobe radius of the white dwarf.

In the following two activities, and in the subsequent questions, we are going to look at the population of magnetic CVs (mCVs), to see if they are consistent with expectations from this simple argument.

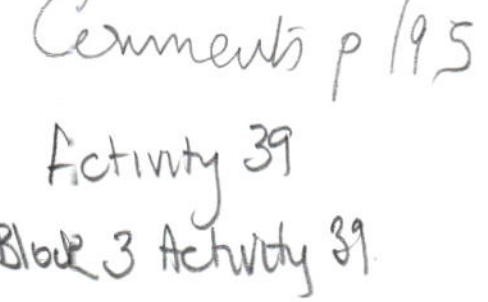

Activity 39 **(30 minutes)**

Orbital period distribution of magnetic CVs: Part I

In this activity you will obtain an up-to-date list of (a) polars and (b) intermediate polars (IPs) with known orbital periods from a catalogue of cataclysmic variables, and save them as electronic files.

Connect to the S381 home page, select Block 3, Activity 39 and follow the instructions on screen.

Keywords: none ■

You can now use the lists you obtained in Activity 39 to generate a period histogram of magnetic CVs.

Activity 40 **(1 hour)**

Orbital period distribution of magnetic CVs: Part II

> Unfortunately, the necessary manipulations described below quite often *fail* and *give erroneous results* if you are using the Microsoft spreadsheet Excel. So, this time unlike in previous spreadsheet activities, you are *advised* to use StarOffice. We are sorry for any inconvenience this may cause you.

Use your spreadsheet program to plot the orbital period distribution of (a) polars, (b) intermediate polars (IPs), and (c) both polars and intermediate polars (mCVs) taken together.

In order to generate an orbital period distribution – or period histogram – you have to divide the period range you wish to consider into small intervals, called period bins. Then count the number of CVs with orbital period in a particular bin, for each bin. This number of systems per bin, plotted against the orbital period of the bin, is the orbital period distribution. Below we describe how to do this with the StarOffice spreadsheet program in detail.

Perform the following steps:

1 Enter the orbital periods into a spreadsheet column. From StarOffice, open the file 'polars' (which you have obtained in Activity 39) as an html file. Select the column that contains the orbital periods (column 8) by dragging the mouse down to the end of the file. Then right-click the selected column, and select 'copy'. Then open a new StarOffice spreadsheet, left-click in the first cell in the first column (column A), then right-click in this cell and select 'paste' from the drop-down menu. The first column of the spreadsheet should now contain the orbital periods of all polars.

2 Convert the orbital periods from days into hours. Click on the first empty cell of the second column (column B). Enter '=A1*24' and press return. (Alternatively, if you chose to put a header or label into the first cell, the first blank cell would be the second cell of column B. In this case you have to enter '=A2*24'). Left-click on the lower-right corner of the frame, hold the mouse button down, and drag it down until it is level with the bottom of the first column. When you release the mouse button, it copies the formula into all cells, and the converted orbital periods appear in column B.

3 Divide the period interval from 1 to 10 hours into bins of equal width 0.25 h (15 min). To do this enter '1' in cell C1 and '1.25' in C2. Then left-click on the lower-right corner of the frame around C1 and C2 and drag the mouse to cell C37. Upon release the column should fill with the period values 1, 1.25, 1.5, 1.75, 2, 2.25, …, 9.75, 10 that separate the period bins.

4 For each period bin, calculate the number of polars with an orbital period in this bin. The function 'frequency' will do that for you. Select cells D1 to D38 (one cell more than in column C). Click on the Insert menu and select Function.... A new dialogue window entitled 'Autopilot: Functions' appears. Find 'Frequency' in the list and double-click on it. Two new fields, entitled 'data' and 'classes' appear on the right-hand side of the dialogue box. Click into the 'data' field, then move the mouse to column B of the spreadsheet. Select column B, i.e. drag the red box down until it surrounds the whole column. The 'data' field should now contain a string, e.g. 'B1:B57' if you have 57 polars in your list. Then click in the 'classes' field and select column C. The 'classes' field should now contain the string 'C1:C37'. Finally, press OK. Column D should now contain the desired number N of systems in the corresponding period bin (e.g. D1 gives the number of polars with periods between 1 and 1.25 h).

5 Plot N in column D versus period in column C. Alternatively, you could also just plot N in a bar diagram. To do the latter, select columns C and D. Then select 'Chart' on the 'Insert' menu. The 'AutoFormat Chart' dialogue box will open. Check 'First column as label' and press 'Next' – this gets you to the second level of the dialogue box. Choose the chart type 'Columns', and be sure that 'Data series in columns' is checked. Then press 'Next'. Choose the 'normal variant', with 'Grid lines Y axis'. Press 'Next'. Enter a suitable chart title (e.g. 'polars'), check 'x-axis title' and enter an x-axis title (e.g. 'Period/h'), check 'y-axis title' and enter a y-axis title (e.g. 'number of systems'). Then press 'Create'. The period histogram should appear in the spreadsheet. Note that the x-axis labels won't look particularly pretty. You may wish to try to resize the chart window (try making the chart wide and short).

Now repeat the above steps 1–5 for intermediate polars.

Finally, copy both the IP and polar data column into one long column in a new spreadsheet, thus creating the complete mCV sample. Then repeat the above steps 1–5 for mCVs.

When you compare your results with the model spreadsheet in the MM resources, please note that the list of systems used for the model spreadsheet is likely to be slightly different from your list – simply because the CV catalogue is continuously updated.

Once you have completed the period histograms, answer the next two questions.

Keywords: **period distribution** ■

Question 70

(a) What main features do you see in the combined period distribution of AM Her stars and intermediate polars taken together?

(b) Now consider the individual period distribution of only polars, and only intermediate polars. The two distributions are different from each other. What is the main difference? ■

Question 71

Derive constraints on the orbital period of polars by using the condition that in a polar the magnetospheric radius R_M must be larger than the Roche lobe radius R_1 of the primary. To simplify the argument, assume that all magnetic CVs have the same surface field strength ($B_{WD} = 10^6$ G), white dwarf mass ($0.8M_{\odot}$) and mass transfer rate ($\dot{M} = 10^{-10}M_{\odot}\,yr^{-1}$). *Hint*: We suggest you tackle the question by breaking it into the following steps:

(a) Start from the condition $R_M > R_1$. Use FKR Equations 6.20 and 6.18 to calculate R_M, and set $R_1 = 0.5 \times a$.

(b) Calculate the magnetospheric radius by inserting the appropriate numbers for the mass transfer rate, mass and magnetic moment of the white dwarf. To determine the magnetic moment, use Nauenberg's mass–radius relation (Block 1, Activity 3) to calculate the radius of the white dwarf.

(c) Use Kepler's law to replace a with the orbital period (you may assume that the total mass of the binary is $1M_{\odot}$).

(d) Solve the inequality for the orbital period (bearing in mind the information in the box 'Manipulating inequalities'). ■

MANIPULATING INEQUALITIES

Inequalities of the form $a < b$ or $c > d$ can be rearranged just as equations, by performing the same operation on both sides of the inequality. The only added complication is that the '$>$' sign becomes a '$<$' (and vice versa) when the operation involves the multiplication or division by a *negative* number or quantity. It is easy to see why: consider as an example the trivial inequality $2 > 1$. Now multiply both sides with (-1). This gives -2 on the left-hand side and -1 on the right-hand side. But obviously $-2 < -1$, so the direction of the inequality had to be reversed.

Equally, if we swap the two sides of an inequality we have to change the direction of the inequality as well: if an inequality reads $a < b$ then we obviously have $b > a$ (not $b < a$), while if $a = b$ we also have $b = a$.

7.2 Magnetically controlled accretion

Magnetically controlled accretion gives rise to a simple and immediately recognizable observational signature because the accretion flow is channelled on to only a small fraction of the stellar surface. Consider the situation depicted in Figure 68 overleaf, an accretion flow through a disc which is disrupted by the dipolar magnetic field of the accreting star at the Alfvén radius. At yet smaller radii the plasma is forced to flow along the field lines. It is clear from the field line pattern shown in Figure 68 that only a small fraction f_{disc} of the surface actually accretes material. A rigorous analysis of the dipolar geometry (as given in FKR – but you do not have to read this in order to progress) shows that that $f_{disc} \approx R_*/R_M$, i.e. the fraction increases when the magnetospheric radius decreases, and when the stellar radius increases.

7.2.1 Accretion columns and curtains

The consequences of this channelled accretion is further explored in your next reading.

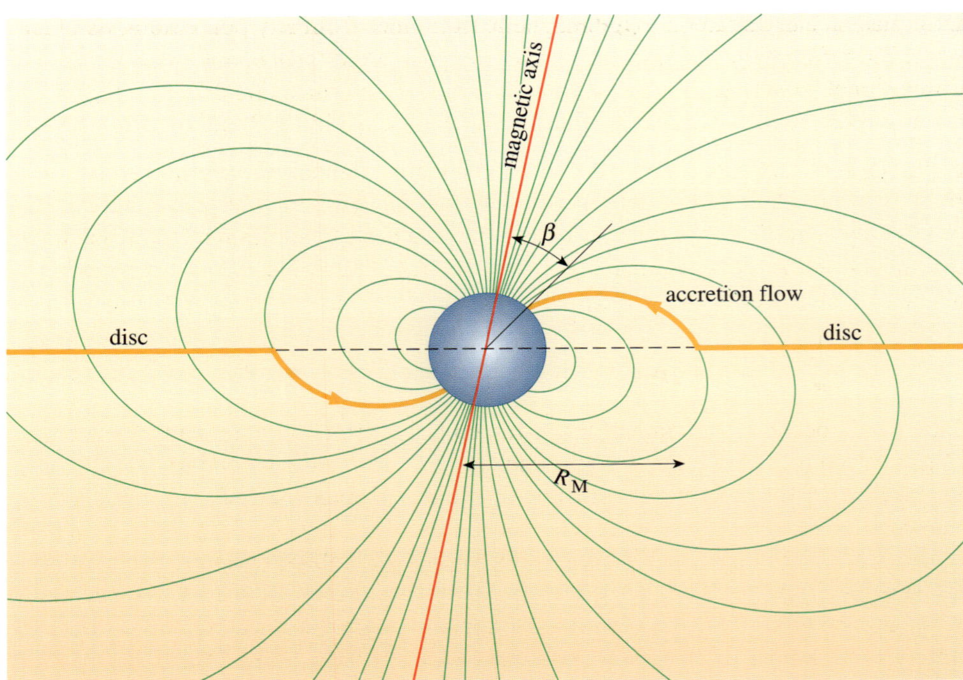

Figure 68 Accretion from a disc on to the magnetic polecaps of a neutron star or white dwarf (this is similar to Figure 6.4 in FKR).

Activity 41 **(30 minutes)**

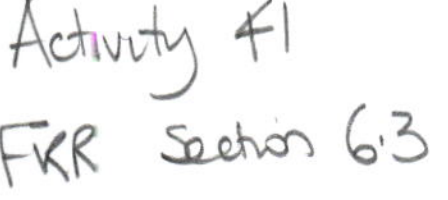

Spin rates

You should now resume reading Section 6.3 of FKR. Read the paragraph containing the above estimate for f_{disc} – FKR Equation 6.22 – on pages 161 and 162 ('Thus in general only a fraction …'). Then jump to page 163 and read the paragraph containing FKR Equation 6.25 and 6.26 (on page 164). From there go to the middle of page 166 and read up to the beginning of the second paragraph on page 168 ('Some progress in understanding …').

As always, you will find assistance for this reading below.

Keywords: **spin modulation**, **co-rotation radius** ■

Let us repeat the various effects on the accretion flow in the presence of a magnetic white dwarf or neutron star.

As you have heard earlier, there are the two different types of magnetic CVs, polars and intermediate polars. In the former, the magnetospheric radius is so large that the mass transfer stream is forced to flow along the field lines very soon after it has left the L_1 point. The white dwarf is in co-rotation with the orbit, i.e. the orbital period and the white dwarf spin period are equal. The stream threads directly on to the field lines and follows them down towards the white dwarf (see Figure 6.7 in FKR and our Figure 50). The incoming material crashes into the surface close to the magnetic pole. The immediately affected area is very small, perhaps only a 10 000th of the surface area. This mode of accretion can be described quantitatively in an idealized way as an **accretion column**, essentially a one-dimensional funnel. The footprint of the accretion column is a hot and luminous, X-ray emitting spot on the surface of the white dwarf. In contrast, in intermediate polars the magnetic field is not strong enough to prevent the formation of an accretion disc altogether, and the white dwarf generally rotates much faster than the orbit. The accretion disc is disrupted at the magnetospheric radius, i.e. the magnetospheric radius defines the inner disc edge.

Disc plasma can thread on to the magnetic field lines from any point along its inner edge. Therefore the ensuing plasma flow along the field lines is not a one-dimensional stream as in the case of polars, but a two-dimensional flow which is usually referred to as **accretion curtain** (see Figure 6.3 in FKR and our Figure 51).

A potential observational signature of magnetically channelled accretion is the rotational modulation of the observed flux when the radiating hot spots rotate in and out of view, similar to the light house effect in pulsars. In addition to a periodicity at the spin period P_{spin}, the optical signal from intermediate polars shows sometimes also a periodicity at the **beat frequency** $\nu_{\text{beat}} = 1/P_{\text{beat}}$, given by

$$\frac{1}{P_{\text{beat}}} = \frac{1}{P_{\text{spin}}} - \frac{1}{P_{\text{orb}}} \tag{76}$$

This would occur when some of the X-ray emission from the polar cap heats an object which is fixed in the binary frame, e.g. the side of the secondary star facing the white dwarf. The heating is then modulated at the **beat period**, P_{beat}. To see how this comes about consider the angular speed of the rotating star, $\omega_{\text{spin}} = 2\pi/P_{\text{spin}}$, and of the orbital motion, $\omega_{\text{orb}} = 2\pi/P_{\text{orb}}$, where P_{orb} is the orbital period (Figure 69). For the sake of simplicity we assume that the heated secondary lights up only when the polar cap is closest to it. By the time the rapidly rotating white dwarf has undergone one full revolution, the secondary has moved on a bit. The pole cap and the secondary line up again after the time P_{beat} has elapsed. During this time the secondary has moved along its orbit by an angle $\delta_1 = \omega_{\text{orb}}P_{\text{beat}}$, while the white dwarf has rotated by an angle $\delta_2 = \delta_1 + 2\pi$.

- ■ Order the periods in Equation 76 for rapidly spinning intermediate polars.
- ❑ In rapidly spinning systems we have $P_{\text{spin}} \ll P_{\text{orb}}$. Looking at Figure 69 it is clear that the beat period is somewhat longer than the spin period, but much shorter than the orbital period. Hence $P_{\text{spin}} < P_{\text{beat}} \ll P_{\text{orb}}$. ■

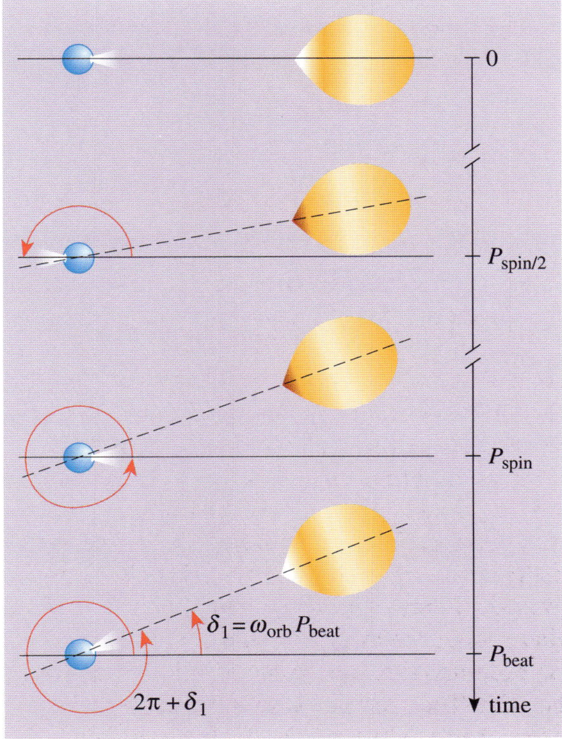

Figure 69 Modulation at the beat period as a result of X-ray heating.

Example 23

Show that the statement $\delta_2 = \delta_1 + 2\pi$ is equivalent to Equation 76.

Solution

Obviously we also have $\delta_2 = \omega_{\text{spin}} P_{\text{beat}}$. Hence $\delta_2 = \delta_1 + 2\pi$ can be written as

$$\omega_{\text{spin}} P_{\text{beat}} = \omega_{\text{orb}} P_{\text{beat}} + 2\pi$$

which is $$\frac{2\pi}{P_{\text{spin}}} P_{\text{beat}} = \frac{2\pi}{P_{\text{orb}}} P_{\text{beat}} + 2\pi$$

Dividing by $2\pi P_{\text{beat}}$ gives

$$\frac{1}{P_{\text{spin}}} = \frac{1}{P_{\text{orb}}} + \frac{1}{P_{\text{beat}}}$$

When solved for $1/P_{\text{beat}}$ this is just Equation 76. ■

A rather intriguing aspect of magnetic accretion is the effect on the spin rate of the accreting star. Consider a slowly spinning magnetic star, surrounded by a viscously spreading accretion torus outside the magnetosphere. When the inner edge of the torus arrives at the magnetospheric radius the plasma has no choice but to abandon its slow inward spiral-in and follow the magnetic field lines instead. The field effectively grabs the plasma and dumps it on to the accreting star. Since angular momentum is conserved the accretor will begin to spin-up. The star continues to spin-up until the specific angular momentum of material on its surface (the spin angular momentum per unit mass) is the same as the specific angular momentum the accreting material brings in. If for some reason the star spins very rapidly to begin with, a most puzzling possibility arises. If the magnetic field sweeping up the plasma from the inner edge of the accretion disc rotates faster than the escape velocity at this point, the material cannot accrete. It is propelled away, out of the system. It is thought that the cataclysmic variable AE Aquarii, where the white dwarf spins once every 33 s, is such a **propeller system**.

Activity 42 **(15 minutes)**

Activity 42
The Energetic Universe MM guide
A propeller in AE Aquarii

A propeller in AE Aquarii

From *The Energetic Universe* MM guide, watch the animation sequence entitled 'A propeller in AE Aquarii'. This is an animated version of Figure 6.8 in FKR. ■

Question 72

(a) What is meant by co-rotation radius?

(b) Calculate the co-rotation radius R_Ω for AE Aquarii. The white dwarf mass is $0.8M_\odot$, the white dwarf spin period is 33 s.

(c) Compare the co-rotation radius with the Roche lobe radius R_1 of the white dwarf. The orbital period is 9.8 h, and the mass ratio is 0.7.

(d) Compare the co-rotation radius with the radius R_* of the white dwarf. ■

Question 73

For a propeller system, is the co-rotation radius R_Ω smaller or larger than the magnetospheric radius R_M? Why? ■

In Question 73 you have found that in propeller systems we have $R_\Omega < R_M$. There are two further inequalities involving characteristic radii that must hold in a propeller system. The white dwarf with radius R_* cannot spin faster than a critical 'break-up' rate, the Keplerian angular speed at radius $R = R_*$. If it did, the then dominant centrifugal forces would induce angular momentum losses that spin-down the white dwarf to a rate below the break-up rate. This fact can be expressed as the requirement $R_* < R_\Omega$. In addition, we need the condition $R_M \lesssim R_1$ (R_1 is the Roche lobe radius of the white dwarf), as otherwise the mass transfer stream would couple to the field immediately upon leaving the inner Lagrangian point, and the system would become a polar. Hence in a propeller system we have the distinct hierarchy of characteristic radii: $R_* < R_\Omega < R_M < R_1$.

7.2.2 Cyclotron radiation

Let us now focus on a completely different, albeit rather important, observational hint for magnetic systems. At the beginning of Section 7 we emphasized that charged particles move on helical trajectories along magnetic field lines. Electrons that move on such trajectories emit a characteristic radiation, called **cyclotron radiation**.

Activity 43

Cyclotron radiation

In Section 6.3 of FKR, read the paragraph containing Equation 6.30 on page 171. It starts on page 170 with ‘ As well as the dynamical …’.

Keywords: **cyclotron radiation**, **polarization**

Electrons that move along magnetic field lines simply due to their random thermal motion generate **thermal cyclotron emission**. In this case most of the energy is emitted at the cyclotron frequency given by FKR Equation 6.30, but some fraction is also emitted at multiples m (m is an integer) of this frequency, the **cyclotron harmonics**. An added complication is that the accretion column in polars is usually opaque to the first few harmonics – only the higher harmonics ($m \gtrsim 5$) escape. Although the observed spectra often show a series of cyclotron humps as in Figure 70,

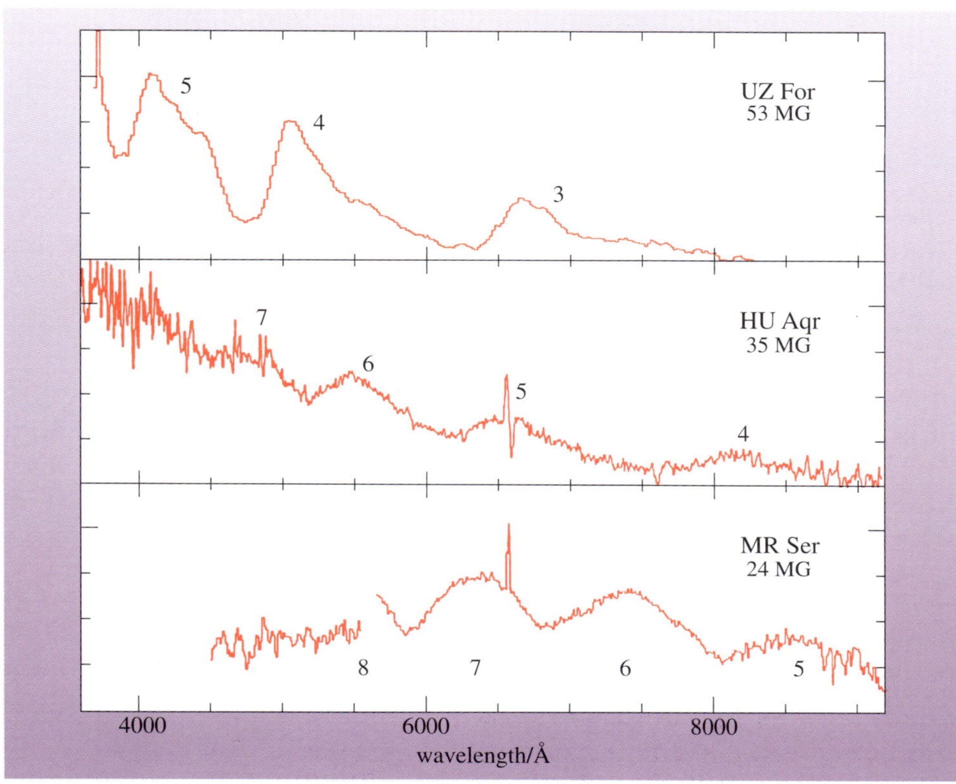

Figure 70 Cyclotron harmonics in the spectra of three different polars. The deduced surface field strength (in MG) and the order m are indicated. (Courtesy of Axel Schwope)

it is usually not clear what order m they belong to, so it is not straightforward to determine the magnetic field strength B from the measured value of ν (FKR Equation 6.30).

The cyclotron frequency, sometimes also called the **Larmor frequency**, is the inverse of the time it takes the electrons to complete one revolution around the magnetic field lines.

Question 74

Referring back to the cyclotron radius (Equation 140 in Block 1), derive the expression for the cyclotron frequency given in FKR Equation 6.30.

Note: FKR Equation 6.30 comes in two versions. In the first version the speed of light c that appears in square brackets is 'active', and therefore cancels with c in the denominator. In this case the cyclotron frequency is

$$\nu_{\text{cyc}} = \frac{eB}{2\pi m_{\text{e}}} \quad \text{(SI)}$$

In the second version the factor c in square brackets is not active, and the cyclotron frequency is

$$\nu_{\text{cyc}} = \frac{eB}{2\pi m_{\text{e}} c} \quad \text{(cgs)}$$

This apparent contradiction is the unfortunate consequence of an ambiguity in the use of units. The formulae in electromagnetism are different for the SI system and the cgs system.

To be consistent with Equation 140 in Block 1 you should stick to SI units when answering this question. This will automatically give the first form of FKR Equation 6.30. ■

Question 75

(a) Calculate the frequency and wavelength of the first two cyclotron harmonics for HU Aqr, using the magnetic field strength quoted in Figure 70.

(b) What range of the electromagnetic spectrum does this correspond to? ■

If the electrons move very fast, in fact with speeds approaching the speed of light, the radiation they emit has a somewhat different character and is called **synchrotron radiation**. You will hear more about this in Block 4 of this course.

But why do these electrons emit radiation at the cyclotron frequency in the first place? This is a property deeply rooted in the physics of electromagnetism. Any charged particle subject to acceleration is liable to emit electromagnetic radiation. If the acceleration involves an oscillatory part with frequency ν, then the emitted radiation has this frequency, too.

Another observational clue to the presence of magnetic fields is the large fraction of polarized light from magnetic CVs, in particular AM Her stars (for an introduction to polarized light see Section 4.9.2 of Block 1). Cyclotron radiation is linearly polarized if viewed in a direction perpendicular to the magnetic field lines, and circularly polarized when viewed in the direction along the field lines. In a circularly polarized electromagnetic wave the plane in which the electric field vibrates is not fixed, but rotates with constant angular velocity around the direction of propagation.

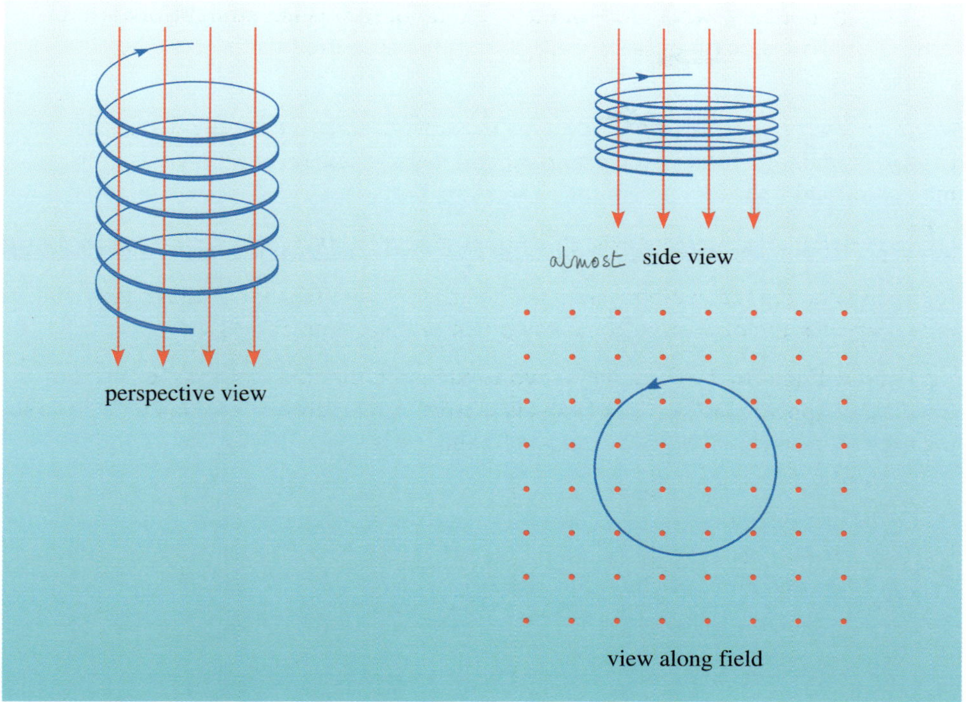

Figure 71 Linear and circular polarization of light emitted from electrons trapped in a magnetic field.

The mode of polarization of cyclotron emission is intuitively clear (see Figure 71). If viewed along field lines the electrons appear to move in circles, giving rise to circular polarization. Viewed from the side the circular orbit is seen 'edge-on' (neglecting the forward motion along the field lines), hence the electron appears to perform a one-dimensional harmonic oscillation.

As the emitting region contains a variety of orientations between the line of sight to the observer and the magnetic field lines, the polarization is never 100%. Also, the radiation will suffer partial depolarization when it interacts with other matter on its way to the observer.

The name game: part 7

Skim-read Section 7 and make a list of all symbols that denote *new* variables and *new* quantities, and note down what they mean. Flag up those symbols that have (unfortunately) multiple uses.

Once you have compiled your list and explained the meaning of all entries, compare it to the one given in Appendix A5 at the end of the Study Guide.

7.3 Summary of Section 7

1 Magnetic fields are frozen in to a perfectly conducting plasma.

2 The importance of a magnetic field for the plasma flow can be assessed by comparing the magnetic pressure

$$P_{mag} = \frac{4\pi}{\mu_0}\frac{B^2}{8\pi}$$

and the ram pressure $P_{ram} = \rho v^2$.

3 The ram pressure is the force exerted on a unit surface area facing the gas stream, due to the linear momentum carried in the flow. If P_{mag} dominates, then the magnetic field determines the fluid velocity, while if P_{ram} dominates, the magnetic field is effectively dragged around by the fluid.

4 The Alfvén radius or magnetospheric radius R_M is the distance from the accreting star where the magnetic pressure equals the ram pressure in the accreting plasma.

5 The simplest global magnetic field pattern is that of a magnetic dipole. The field strength of a dipolar field varies roughly as

$$B \approx \frac{\mu}{r^3}$$

with distance r from the star. The quantity μ is the magnetic dipole moment.

6 The conservation of magnetic flux ($\propto BR^2$) implies that compact stars can have very strong surface magnetic field strengths.

7 A magnetically controlled accretion flow is channelled on to only a small fraction $\sim R_*/R_M$ of the accretor's surface.

8 In the disc-less polars the white dwarf is in co-rotation with the orbit, and the mass transfer stream threads directly on to the field lines. This mode of accretion can be described as an accretion column. The footprint of the accretion column is a hot and luminous, X-ray emitting spot on the surface of the white dwarf. Typical surface field strengths are 10 to 100 MG; μ is of order $10^{34}\,\mathrm{G\,cm^3}$.

9 In intermediate polars the white dwarf always rotates faster than the orbit. The accretion disc is disrupted at the magnetospheric radius, and an accretion curtain forms.

10 The observed flux from compact stars with magnetically channelled accretion may vary on the spin period P_{spin} or the orbital period P_{orb} when the radiating hot spots rotate in and out of view. The signal may also vary on the beat period P_{beat},

$$\frac{1}{P_{beat}} = \frac{1}{P_{spin}} - \frac{1}{P_{orb}}$$

if, for example, an object that is fixed in the binary frame is periodically heated on the spin period.

11 Propeller systems eject the accreting plasma as the co-rotation radius is smaller than the magnetospheric radius. The propeller effect spins-down the rapidly rotating primary star.

12 Magnetic CVs emit thermal cyclotron radiation, and a significant fraction of the radiation is polarized.

8 DISCS IN TURMOIL: OUTBURSTS AND LIMIT CYCLES

In this final major section of Block 3 we return to accretion discs around non-magnetic or weakly magnetic stars. So far, we have essentially ignored time-dependent phenomena in the accretion disc itself. This was largely driven by our desire to simplify the all too complex hydrodynamic problem of the flow of astrophysical fluids. The main simplification was the assumption of time-independence, leading to the theory of steady-state accretion. We found that many accreting systems are well described in this way. Plenty of observational evidence broadly confirms the main findings of steady-state theory.

The obvious drawback is that the theory of steady-state accretion cannot deal with time-variability, including such spectacular phenomena as dwarf novae (Figure 11) and soft X-ray transient outbursts (Figure 16). There is also a second, more subtle drawback. By observing purely steady-state phenomena we fail to discover the origin and magnitude of the most uncertain physical quantity involved in the theory of accretion, the viscosity. For, as you will remember, the viscosity does not appear in the steady-state expression for the temperature and the radiant flux emerging from the disc surface (see, for example, FKR Equations 5.20, 5.21 and 5.43). Refer back to the answer for Question 48 if you are unsure why this implies that we cannot discover the magnitude of viscosity.

In principle, there are now various ways to proceed. We could write down disc equations that encapsulate the full time-dependent physics, and then set out to find a solution of these equations – Σ as a function of R and t, say – by whatever means. In fact, non-linear, higher-order partial differential equations like these disc equations can indeed be solved numerically, with sophisticated computer codes. Here we can only summarize the basic procedure of such a numerical solution. You do not have to remember these details for a successful completion of the course, so there is no need to worry if some of the terms we use are unfamiliar to you, or if you feel you would need more explanation.

For the purpose of a numerical solution the equations have to be rewritten in an approximate way as difference equations, i.e. equations where differentials of the form dR or dt are replaced by small differences ΔR and Δt between the values of R and t at neighbouring points in space and time. The resulting new system of equations is solved for $\Sigma(R, t)$ on this discrete grid of points in space and time. The technique used to find a solution can be very complicated and tedious. In many cases it involves an iteration, reminiscent of the procedure you heard about for Doppler tomography. A guessed or initial trial solution for the first time grid point (or points) is fed into the system of equations. The trial solution usually fails to solve the equations, and from the degree of failure a correction can be calculated. When applied to the first estimate a new, better trial solution is produced. This is repeated until the suggested solution solves the equations within a suitably chosen numerical accuracy. Then the iteration is said to have reached convergence, and the next point on the time grid can be considered. Quite often there are problems in achieving convergence, and even if it is achieved, it is not easy to see if the accepted accuracy is good enough to guarantee that the numerical solution of the difference equations is also a solution of the original differential equation.

We could now simply present you with the numerical solution found for time-dependent accretion in a thin disc, but this would not provide you with any insight into the physics of the problem. You would see that the calculations do indeed reproduce a time-variability of the disc luminosity which is reminiscent of dwarf nova outbursts, but clearly we want to gain an understanding of *why* such outbursts occur, and identify the physical mechanisms determining the outburst characteristics.

The approach we take is to investigate the steady-state disc solutions and search for *instabilities*. Are there instabilities – not described by steady-state theory – which might prevent real accretion discs from achieving a steady state in the first place? If so, on what timescale do these instabilities grow, and what is the likely outcome of a system subject to the instability? The advantage of this approach is that we will be able to describe and understand certain time-dependent phenomena by considering transitions between different steady states.

Do not worry if this sounds still rather mysterious and theoretical to you. Just bear with us and follow the reasoning below which will eventually culminate in the main multimedia activity of this block.

You will also find that this section will revise many of the physical concepts we discussed earlier in this block, because the analysis starts from the steady-state model we have developed in Sections 4 and 5.

8.1 Thermal instability

We now set out and specify exactly what we mean by time-dependent accretion. Time-dependence can occur on a number of very different timescales, and it is worthwhile sitting back and considering exactly what timescale we are most interested in. For instance, you are already familiar with the viscous time in Section 4.1.4, and you also know that another characteristic timescale is the time it takes matter in the accretion disc to complete one revolution. This is the so-called dynamical time.

8.1.1 Timescales

In the following activity, FKR introduces yet another timescale, and clarifies the relation between the three timescales. FKR uses the symbol $\mathcal{M}$ to denote the **Mach number** of the flow, i.e. the azimuthal (Keplerian) velocity in units of the sound speed,

$$\mathcal{M} = \frac{v_\phi}{c_s} \tag{77}$$

Activity 44 (30 minutes)

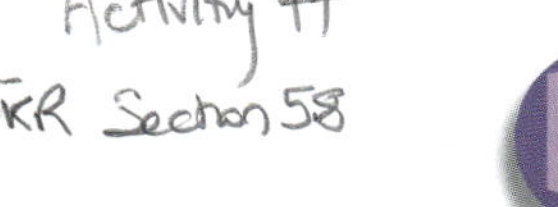

The hierarchy of timescales

Read Section 5.8 of FKR up to the end of the penultimate paragraph on page 113 ('… an 'equilibrium' solution can be found').

Keywords: **viscous timescale**, **dynamical timescale**, **thermal timescale**, **thermal instability**, **cooling rate**, **heating rate** ■

Let us summarize the distinct hierarchy of timescales as discussed in FKR. The shortest time is the dynamical time, the time it takes disc particles to complete one Kepler orbit around the accreting star. For typical CV discs the dynamical time is of order minutes. This timescale is up to a factor of order unity the same as the free-fall time (e.g. Section 1.2.3 of Block 2).

The free-fall or dynamical time is also the characteristic time for re-establishing hydrostatic equilibrium in the direction perpendicular to the disc plane, should this equilibrium be perturbed. FKR appeals to FKR Equation 5.32 – an expression that was not part of your reading – to show this. Consider Example 24 as an alternative way to arrive at the same result.

CHARACTERISTIC TIMESCALES

Quite often you find statements to the effect that a certain time or timescale is *characteristic* of a certain physical effect. For instance, the dynamical time is the characteristic time to establish hydrostatic equilibrium. What this really means is that the *actual* time it takes for the process to occur is *of order* of that characteristic time. Depending on the situation the actual time could easily be half or twice as long as the characteristic timescale. It could also simply mean that it takes the characteristic time for any *significant changes* to occur. An example is the radioactive decay process. The decay rate decreases exponentially with time (Figure 72), so strictly speaking the process of decay never finishes. But significant changes – the decay of 50% of the nuclei – occur within the radioactive half-life.

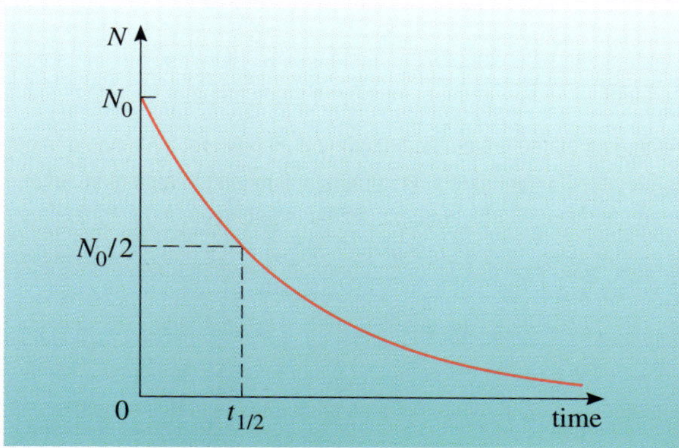

Figure 72 Characteristic timescale for exponential decay.

Example 24

Use the expression for the disc thickness (FKR Equation 5.25) to show that the time t_z (defined in FKR Equation 5.64) is the same as the dynamical time (FKR Equation 5.63).

Solution

FKR Equation 5.25 reads

$$H \cong c_\mathrm{s}\left(\frac{R}{GM}\right)^{1/2} R$$

Hence $t_z = H/c_\mathrm{s}$ (FKR Equation 5.64) is

$$t_z = \frac{H}{c_\mathrm{s}} \cong \left(\frac{R}{GM}\right)^{1/2} R = \left(\frac{R^3}{GM}\right)^{1/2}$$

Now as the Keplerian angular speed is just $\Omega = \Omega_\mathrm{K} = (GM/R^3)^{1/2}$ (Equation 29) we have $t_z \sim 1/\Omega_\mathrm{K}$, which is exactly the same as for the dynamical time, $t_\phi \sim 1/\Omega_\mathrm{K}$ (FKR Equation 5.63). ■

The (local) thermal time measures how long it takes to generate the thermal energy content of a disc annulus by viscous dissipation at the current rate. This is similar in spirit to the Kelvin–Helmholtz time in stars, $t_{\rm KH} = GM_*^2/R_*L_*$, which is the time it takes to remove the heat content (thermal energy) $\propto GM_*^2/R_*$ of a star with mass M_* and radius R_* if the star radiates at luminosity L_*. The significance of the thermal time is that it indicates the characteristic timescale to re-establish **thermal equilibrium** should this be perturbed. In the state of thermal equilibrium any local energy losses from the disc, e.g. via radiation, are exactly balanced by the energy which viscous dissipation generates at this disc annulus.

The thermal time in discs turns out to be longer than the dynamical time by a factor $1/\alpha$. To calculate this FKR starts by noting that the heat content of a gas per unit volume is $\propto \rho c_{\rm s}^2$. This is because the heat content of N gas particles with average mass $\mu m_{\rm H}$ which populate a volume V at temperature T is $\propto NkT$. Here μ is the mean molecular weight in units of the hydrogen mass (not the magnetic moment as in Section 7). The heat content per volume then is proportional to

$$\frac{NkT}{V} = \frac{(M/\mu m_{\rm H})}{V}kT = \frac{M}{V\mu m_{\rm H}}kT = \frac{\rho}{\mu m_{\rm H}}kT$$

where we have used the identity $N = M/(\mu m_{\rm H})$ and $\rho = M/V$. But $kT/(\mu m_{\rm H})$ is just the square of the isothermal sound speed $c_{\rm s}^2$ (Equations 25 and 26). As usual, the transition from a volume density to a surface density is achieved by integrating over the vertical direction, i.e. by replacing ρ with Σ.

FKR relates both the dynamical time and the thermal time to a time you are already familiar with, the viscous time $t_{\rm visc}$ (see Equation 53). FKR finally finds

$$t_\phi \approx t_z \approx \alpha t_{\rm th} \approx \alpha\left(\frac{H}{R}\right)^2 t_{\rm visc} \qquad (78)$$

which shows that for $\alpha < 1$ the dynamical time is shorter than the thermal time, while both are much shorter than the viscous time (since $H/R \ll 1$).

■ What are typical values of the three timescales at the outer edge of a Shakura–Sunyaev disc in a binary with a few hours orbital period? Assume $\alpha = 0.1$.

❑ The dynamical time is the Kepler period of disc particles. According to Kepler's law, the period scales with the distance from the central object as $P \propto R^{3/2}$. Hence the dynamical time in the outer disc is shorter than the binary period, as the disc is smaller than the binary itself. Let's say $t_z = 1$ h. Then the thermal time is 10 h, while the viscous time is much longer. With $H/R \approx 0.01$ we have $t_{\rm visc} \approx 10\,{\rm h}/0.01^2 = 10^5$ h, i.e. more than 10 years. ■

This information helps a great deal in analysing a non-equilibrium situation where locally, in a particular disc annulus, the rate of disc heating is out of step with the rate of disc cooling. Clearly, in the case of such an imbalance the disc must react, e.g. by becoming hotter when heating dominates over cooling. More specifically, the disc will heat up locally on the characteristic timescale associated with a perturbation of thermal equilibrium – the thermal time. If as a result of the disc's immediate reaction to a small perturbation the heating rate is even more out of step with the cooling rate the perturbation will grow and the disc will continue to heat up, on a thermal timescale. It is said that the disc is subject to a thermal instability, and the growth time of the instability is the thermal time.

STABILITY, INSTABILITY AND STABILITY ANALYSIS

The meaning of a statement like '*a physical system is stable*' can be illustrated with a simple example: the pendulum. Consider an idealized pendulum on Earth, consisting of a weight with mass m that is attached to one end of an essentially massless rod with length l (Figure 73). The other end of the rod is fixed to a horizontal axis. The pendulum is allowed to rotate freely around this axis in a fixed vertical plane. Assume that the only forces acting on the pendulum are the gravity $m\boldsymbol{g}$ on the weight ($\boldsymbol{g}$ is the gravitational acceleration on the surface of the Earth), and a certain amount of friction, so that the swinging pendulum would eventually come to rest.

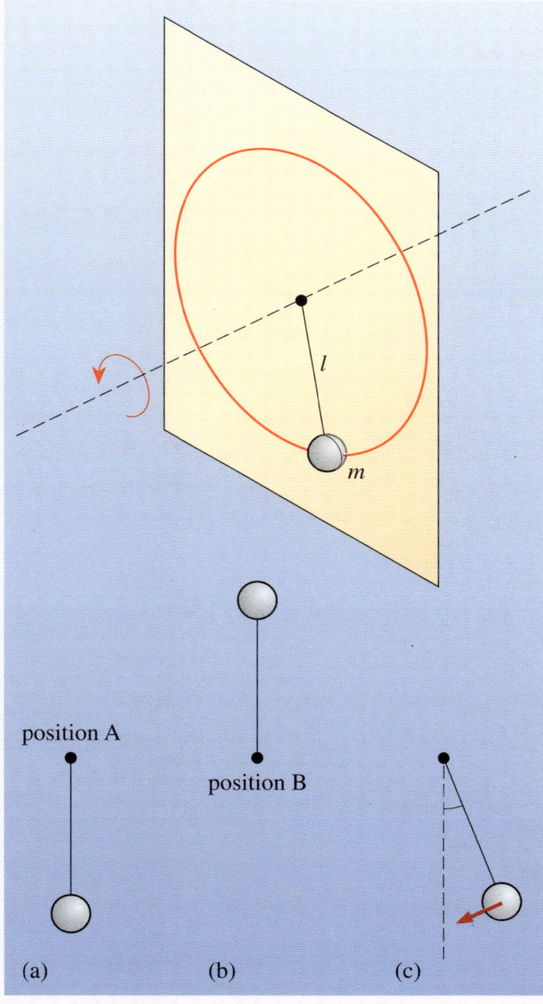

Figure 73 Equilibrium positions of a pendulum.

There are exactly two equilibrium positions of the pendulum, i.e. positions where the weight is at rest and stays at rest (at least in principle). The first such position (A) is when the weight is at the lowest point (Figure 73a). The second position (B) is when the weight is at the highest point (Figure 73b). Your own experience certainly tells you that in practice the pendulum would stay at rest only in equilibrium A. Even if you managed to halt the pendulum in the upside-down position B, the slightest vibration or air breeze would perturb the equilibrium. The pendulum would inevitably swing round and eventually end up in the equilibrium position A. This is the basis for saying that equilibrium A is *stable*, while equilibrium B is *unstable*.

More formally, a physical system is said to be stable, or in a stable state, if a small perturbation, after it has been applied to this state, will die away by itself, thus re-establishing the original state. This is the case for the pendulum in position A. If we displace the weight by a small angle and then release it (Figure 73c), there is a net force on the pendulum that causes an acceleration towards position A. The pendulum would oscillate around the equilibrium position A for a short while, but due to friction it will eventually come to rest at position A.

Conversely, a physical system is said to be unstable, or in an unstable state, if a small perturbation, after it has been applied to this state, will grow by itself, and lead the system further away from the original state. This is the case for the pendulum in position B. If we displace the weight by a small angle and then release it, there is a net force on the pendulum that causes an acceleration *away* from position B.

Although the stability behaviour is intuitively clear for the pendulum, there are plenty of physical systems where this is not the case – one example is accretion discs. We have discussed steady-state accretion discs at length, but so far we know nothing about whether a disc in such a steady state is *stable*. This is a rather important point, because in case of instability we would not expect to see a disc in this state – it is so easy to destroy it.

A formal stability analysis can show if a physical system is in a stable state or not. In such a stability analysis one considers small perturbations to the state – like the small displacements applied to the pendulum – and *calculates* how the system would react to it. *The system is stable if it opposes the initial perturbation, the system is unstable if it amplifies the initial perturbation.*

As the dynamical time is short compared to this growth time the disc adjusts its vertical structure effectively instantaneously, and is in hydrostatic equilibrium at all times during the growth of the thermal instability. But despite the readjustment the local surface density does not change.

■ Why?

❑ The only way Σ could change is through viscous transport of disc mass from or to neighbouring disc annuli. But this occurs on the viscous timescale, which we know is long compared to the thermal time. ■

8.1.2 Heating and cooling

A thermal instability will be at the heart of our model for dwarf nova and soft X-ray transient outbursts, so it is a good idea to put the conditions for it in a more quantitative form. FKR denotes the local disc heating and cooling rate per unit volume as Q^+ and Q^-.

To be precise, Q^+ denotes the heat energy generated per unit volume per unit time (and is measured in erg s^{-1} cm^{-3}). You have met Q^+ already in Section 5.2 above, and in the corresponding reading, FKR Section 5.4 (you might wish to revise this now). We know the viscous dissipation rate D per *unit surface area* from our earlier discussion (Equation 37), so the rate per *unit volume* is

$$Q^+ \approx \frac{D(R)}{H} \tag{79}$$

It is instructive to look into how this is obtained. This revises the thinking behind the use of vertically integrated quantities. First, on dimensional grounds it is clear that the dissipation rate per unit area must be converted into a dissipation rate per unit volume through division by a length. The scale height H of the annulus is a characteristic length of the problem, so Equation 79 seems reasonable enough. The proper way to derive it is as follows. The dissipation rate per unit face area of the disc annulus is defined as the vertical integral of Q^+:

$$\int_0^\infty Q^+(z)\,\mathrm{d}z = D(R) \tag{FKR 5.39}$$

Now the fact is that we do not know how the viscous dissipation is distributed with z. This depends on the detailed physics of the mechanism providing viscosity. A good guess is probably that most of the dissipation occurs close to the mid-plane, where the temperature is highest. But how concentrated to the mid-plane is it? This uncertainty is just one of the many uncertainties inherent in even the advanced and detailed models of accretion discs which researchers compute and work with today.

■ In stars, is there a similar uncertainty in the distribution of the heating rate?

❑ No. Stars are heated by nuclear fusion. The fusion rates are known and drop steeply with decreasing temperature. Hence heating in stars by nuclear energy generation is confined to the core regions of stars where it is hottest. ■

So all we can derive – and all we need to know for what is to follow – is an average or representative value $\langle Q^+ \rangle$ for Q^+. Hence we have

$$D(R) = \int_0^\infty Q^+(z)\,\mathrm{d}z = \langle Q^+ \rangle H$$

The second equality can be understood as a *definition* of $\langle Q^+ \rangle$, i.e.

$$\langle Q^+ \rangle = \frac{D(R)}{H}$$

It is a bit cumbersome to use the notation $\langle Q^+ \rangle$ all the time, so purely for convenience, we (and FKR) drop it, and simply write Q^+ instead of $\langle Q^+ \rangle$. This reproduces Equation 79. But keep in mind the fact that we are really talking about an average value.

Now let us consider the cooling rate. Without knowing it you have met Q^- already in FKR Section 5.4. The rate Q^- did not appear explicitly as it was always *assumed* that the disc is in equilibrium, and hence that $Q^- \equiv Q^+$. The innocent looking unnumbered equation in FKR on page 89 between FKR Equations 5.38 and 5.39,

$$\frac{\partial F}{\partial z} = Q^+$$

expresses just that (see also Section 5.2). Here the left-hand side *is* the cooling rate Q^-; it *is not* a definition for Q^+! The equality expresses the balance between heating and cooling. So this should really read

$$\frac{\partial F}{\partial z} = Q^- \tag{80}$$

and $$Q^- = Q^+$$

Recall that the quantity $F(z)$ is the flux of energy in the z-direction at height z above the disc mid-plane, measured in units such as erg s^{-1} cm^{-2}. This is analogous to the flux in the stellar interior, $F(r) = L_r/4\pi r^2$ in the radial direction. In other words, F measures the amount of energy transported per unit time through a surface of unit area. If this energy transport is mediated by the diffusion of photons, i.e. if the disc is radiative, then there is a well-known expression for F in terms of the temperature gradient and the opacity κ,

$$F(z) = \frac{-16\sigma T^3}{3\kappa\rho}\frac{\partial T}{\partial z} \tag{81}$$

This is also FKR Equation 5.37, the only difference being that we have dropped the subscript 'R' of the Rosseland mean opacity κ_R. You have met this expression describing the diffusion of radiation also in Block 2 (Section 2.2.2).

Example 25

Using arguments similar to those we used to calculate an average value of Q^+, show how FKR arrives at

$$Q^- = \frac{dF}{dz} \approx \frac{4\sigma T_c^4}{3\kappa\rho H^2} \tag{82}$$

for an estimate of an average value of Q^-. (See the penultimate equation on page 114 of FKR, where the factor 4/3 has been neglected.)

Solution

First we note that FKR apparently does not distinguish between d/dz and the partial derivative $\partial/\partial z$. This is a bit sloppy but justified as we are exclusively dealing with the disc structure perpendicular to the plane, at a fixed disc annulus. So all derivatives with respect to z are performed with the implicit understanding that R = constant. Second, using the chain rule we note that $d(T^4)/dz = d(T^4)/dT \times dT/dz = 4T^3(dT/dz)$. So Equation 81 can be rewritten as

$$F(z) = \frac{-16\sigma T^3}{3\kappa\rho}\frac{\partial T}{\partial z} = \frac{-4\sigma}{3\kappa\rho}4T^3\frac{\partial T}{\partial z} = \frac{-4\sigma}{3\kappa\rho}\frac{\partial (T^4)}{\partial z} \tag{83}$$

We now write the definition of Q^-, Equation 80, in integral form:

$$\int_0^H \frac{\partial F(z)}{\partial z}\,dz = \int_0^H Q^-(z)\,dz$$

Evaluating the first integral, and defining $\langle Q^- \rangle$ in an analogous way as $\langle Q^+ \rangle$, this becomes

$$F(H) - F(0) = \langle Q^- \rangle H \quad \text{or} \quad \langle Q^- \rangle = \frac{F(H) - F(0)}{H} = \frac{F(H)}{H}$$

In the last step we used $F(0) = 0$, as for reasons of continuity the temperature gradient in the mid-plane must be zero (this is analogous to saying that $L_r = 0$ in the centre of a star). We now need to estimate $F(H)$. To do so we first replace the gradient dT^4/dz by a suitable ratio of differences, $\Delta T^4/\Delta z$. Specifically, we use the values of T at $z = H$ and the mid-plane temperature to obtain the admittedly crude estimate

$$\frac{d(T^4)}{dz} \approx \frac{T(H)^4 - T(0)^4}{H}$$

The disc mid-plane temperature $T(0) = T_c$ taken to the fourth power is much larger than the surface temperature $T(H)$ taken to the fourth power, so that the latter can be neglected:

$$\frac{d(T^4)}{dz} \approx \frac{T(H)^4 - T(0)^4}{H} \approx -\frac{T(0)^4}{H}$$

Note the minus sign, as the temperature drops away from the mid-plane. Using this in Equation 83 gives

$$F(H) \approx \frac{4\sigma}{3\kappa\rho}\frac{T_c^4}{H}$$

where again $\kappa\rho$ has to be seen as a suitable mean value of $\kappa(z)\rho(z)$. Inserting this last expression for $F(H)$ into $\langle Q^- \rangle = F(H)/H$ from above we have

$$Q^- = \frac{4\sigma T_c^4}{3\kappa\rho H}\bigg/ H = \frac{4\sigma T_c^4}{3\kappa\rho H^2}$$

which is Equation 82. ■

Question 76

Show that a Kramers opacity $\kappa \propto \rho T_c^{-7/2}$ gives the temperature dependence $T_c^{15/2}$ for Q^- if the surface density is constant. Use Equation 82 for Q^-, and introduce Σ into this equation. ■

We now return to the question of thermal stability of an accretion disc. From what has been said above it is clear that the disc is unstable if a temperature perturbation keeps growing once applied. This will be the case if, as a result of a slight increase dT_c of the central temperature, the heating rate increases relative to the cooling rate, i.e. if

$$\frac{d(Q^+ - Q^-)}{dT_c} > 0 \tag{84}$$

so that the temperature grows even further, as there is more heating than cooling.

Question 77

Convince yourself that the criterion represented by Equation 84 expresses instability also in the case when the initial perturbation goes to slightly cooler temperature. ■

8.2 Time-dependent disc equations

We are now in a position to generalize our model description of steady-state, optically thick, geometrically thin accretion discs to viscously evolving discs. We have seen that the dynamical and thermal time are both much shorter than the viscous time. It will simplify our model a great deal if we assume that the viscously evolving disc is always in hydrostatic *and* thermal equilibrium. We shall relax the assumption of thermal equilibrium shortly.

Activity 45

Viscous disc evolution

Read page 115 of FKR, starting with the second paragraph ('Let us now consider …').

Do not waste time by trying to follow the derivation of FKR Equation 5.73 – this equation is not really needed in the following.

Keywords: none ■

Question 78

Compare the set of time-dependent disc equations (FKR Equation 5.74) to the set of time-independent disc equations (FKR Equation 5.41). Identify the differences, and explain where they come from. ■

We now turn our attention to one of the new equations in the set FKR Equation 5.74, the diffusion equation no. 7. In fact, this is our old friend, Equation 45,

$$\frac{\partial \Sigma}{\partial t} = \frac{3}{R}\frac{\partial}{\partial R}\left[R^{1/2}\frac{\partial}{\partial R}(\nu\Sigma R^{1/2})\right]$$

which you have met already, e.g. in Activity 19. Previously we considered the diffusion process expressed by Equation 45 for the special case when the viscosity ν is constant. We know by now that rather successful disc models make extensive use of the α-viscosity, $\nu = \alpha c_s H$ (see Equation 39). This α-viscosity is far from constant since the sound speed c_s is a function of temperature and therefore radius R, and the disc height H is a function of R as well.

It is possible to show that in most cases the set FKR Equation 5.74 allows one to write the viscosity as a function of only two main variables. The first is the surface density Σ, which can be thought of as the main unknown variable one wishes to calculate by solving the system FKR 5.74. The second is the independent variable R; all unknown functions are to be calculated as a function of radius, e.g. $\Sigma(R)$. Formally, we have $\nu = \nu(\Sigma, R)$. This is important as it helps us in solving the diffusion equation.

Activity 46

Viscous instability

Time for another mini-dose of FKR Section 5.8. Start reading from about the middle of page 116, 'the set (5.74) fixes …', and read up to the end of the second paragraph on page 117 ('…, and vice versa.').

Be warned that FKR introduces the symbol μ for yet another quantity, the product $\nu\Sigma$. This has nothing to do with the mean molecular weight, or the magnetic moment, or the magnetic permeability. It is just a variable used to denote the product $\nu\Sigma$. Confusing though it may seem, it simply shows once again that the Greek alphabet does not have enough letters to satisfy an astrophysicist's needs.

Also be warned that this and the following reading will be rather hard going. But by now you certainly know better than to panic! There is no need to dwell on the equations for long – we shall discuss them in detail below. Try to focus on what it all means in physical terms.

Keywords: **viscous instability** ■

In this last reading FKR performed a number of manipulations on the diffusion equation, with the aim of deriving a slightly modified diffusion equation for the perturbation $\Delta\mu$. Before we look in detail at these manipulations we should be clear about why this is going to help us understand the fate of the accretion disc. FKR starts out from an apparently peaceful, steady accretion disc. FKR then tests how peaceful this disc really is by pushing it a little at one point, i.e. by suddenly dumping a little bit of additional mass on one disc ring. This increases the surface density Σ_0 at this disc ring by a small amount $\Delta\Sigma$. And then FKR sits back and watches how the disc reacts to this change. The diffusion equation comes in because the further fate of $\Delta\Sigma$ (or, to be precise, the fate of its close cousin, $\Delta\mu$) is a solution of this equation. But as we *know* how the solutions of the diffusion equation behave we then also *know* the behaviour of the perturbation, and hence can predict the fate of the disc.

If the magical coefficient $\partial\mu/\partial\Sigma$ in the diffusion equation for $\Delta\mu$ (last equation on page 116 of FKR) is positive, the solutions are just like the one you have seen in Activity 19. Any perturbation of the density profile *diffuses* or decays away on a diffusion timescale. The disc therefore is stable against such perturbations, the disc is viscously stable.

■ As $\mu = \nu\Sigma$, why is the derivative $\partial\mu/\partial\Sigma$ not just equal to ν?

❑ Because ν is a function of Σ (and other variables) itself. So one has to apply the product rule to calculate the derivative:

$$\frac{\partial\mu}{\partial\Sigma} = \frac{\partial(\nu\Sigma)}{\partial\Sigma} = \nu\frac{\partial\Sigma}{\partial\Sigma} + \Sigma\frac{\partial\nu}{\partial\Sigma} = \nu + \Sigma\frac{\partial\nu}{\partial\Sigma}$$ ■

If on the other hand $\partial\mu/\partial\Sigma$ is negative (for the purposes of this analysis the coefficient is just a constant – but it can be either positive or negative), the solutions behave altogether differently. They *grow*. The perturbation grows, causing the affected disc ring to accumulate mass, while neighbouring disc rings are depleted in mass (see Figure 74 overleaf). The disc is breaking up into detached rings! Clearly, in this case the disc is viscously unstable.

To summarize: the disc is viscously stable if

$$\frac{\partial\mu}{\partial\Sigma} > 0 \tag{85}$$

We are now engaging in a rather detailed mathematical derivation of the modified diffusion equation for the perturbation $\Delta\mu$. This might look frightening at first sight, but if you go through the following carefully step by step you will learn a great deal about how issues of *stability* are investigated. It also shows you how to perform very basic manipulations on rather complex looking equations with nested partial differentials.

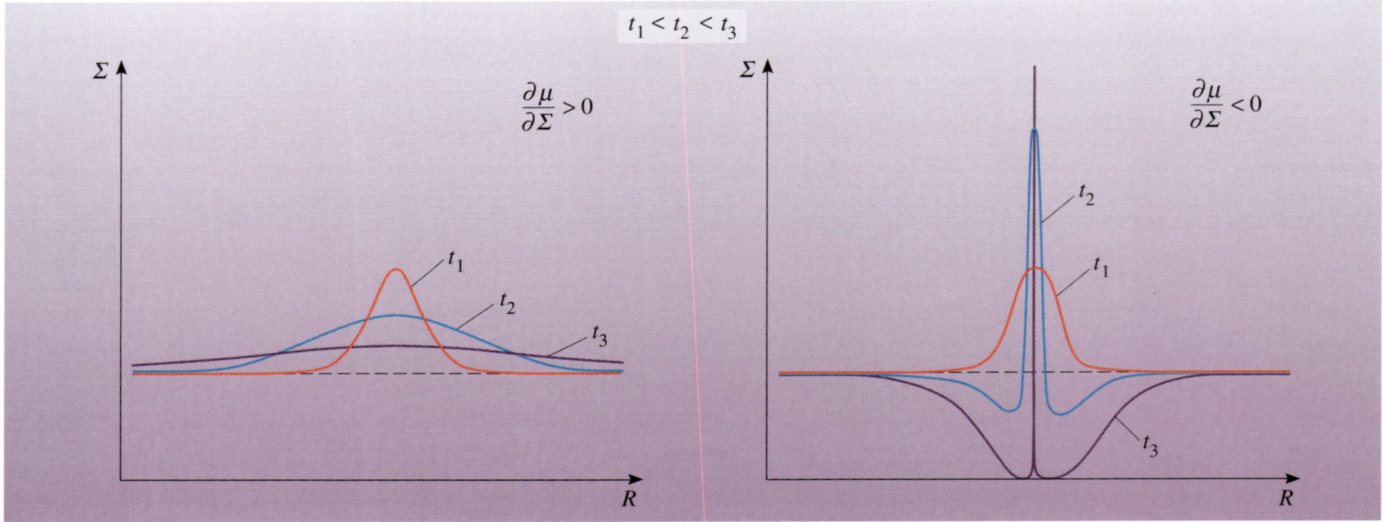

Figure 74 Behaviour of a surface density perturbation for (left) viscously stable and (right) viscously unstable discs.

Our starting point is a steady-state disc, represented by the time-independent surface density profile $\Sigma_0(R)$. In practice this means that if $\Sigma_0(R)$ is inserted for Σ in the diffusion equation, Equation 45, both sides of the equation will be identical to zero,

$$\frac{\partial \Sigma_0}{\partial t} = \frac{3}{R}\frac{\partial}{\partial R}\left[R^{1/2}\frac{\partial}{\partial R}(\nu_0 \Sigma_0 R^{1/2})\right] = 0 \tag{86}$$

Here ν_0 denotes the corresponding viscosity $\nu_0 = \nu(\Sigma_0, R)$ in the steady-state disc. Next we consider the new, perturbed surface density profile $\Sigma_1(R) = \Sigma_0(R) + \Delta\Sigma(R)$. (Initially the perturbation is arranged such that it is localized, centred on a certain disc ring with radius R, as shown in Figure 74.) Clearly, this new $\Sigma_1(R)$ has to obey the diffusion equation as well,

$$\frac{\partial \Sigma_1}{\partial t} = \frac{3}{R}\frac{\partial}{\partial R}\left[R^{1/2}\frac{\partial}{\partial R}(\nu_1 \Sigma_1 R^{1/2})\right]$$

Using the notation 'μ' for the product '$\nu\Sigma$' we have $\mu_1 = \nu_1\Sigma_1$. Just as for Σ we write this as the sum of the original, unperturbed contribution, plus the perturbation, i.e. as $\mu_1 = \mu_0 + \Delta\mu$, where $\mu_0 = \nu_0\Sigma_0$. Then the above diffusion equation for Σ_1 can be written as

$$\begin{aligned}\frac{\partial(\Sigma_0 + \Delta\Sigma)}{\partial t} &= \frac{3}{R}\frac{\partial}{\partial R}\left\{R^{1/2}\frac{\partial}{\partial R}[(\mu_0 + \Delta\mu)R^{1/2}]\right\}\\ &= \frac{3}{R}\frac{\partial}{\partial R}\left[R^{1/2}\frac{\partial}{\partial R}(\mu_0 R^{1/2} + \Delta\mu R^{1/2})\right]\end{aligned}$$

In the last step we have carried out the multiplication within the square bracket. Now using the sum rule for differentiation on the left-hand side and for the inner derivative on the right-hand side of the equation we have

$$\frac{\partial \Sigma_0}{\partial t} + \frac{\partial(\Delta\Sigma)}{\partial t} = \frac{3}{R}\frac{\partial}{\partial R}\left[R^{1/2}\frac{\partial}{\partial R}(\mu_0 R^{1/2}) + R^{1/2}\frac{\partial}{\partial R}(\Delta\mu R^{1/2})\right]$$

Using the sum rule again on the right-hand side gives

$$\frac{\partial \Sigma_0}{\partial t} + \frac{\partial(\Delta\Sigma)}{\partial t} = \frac{3}{R}\frac{\partial}{\partial R}\left[R^{1/2}\frac{\partial}{\partial R}(\mu_0 R^{1/2})\right] + \frac{3}{R}\frac{\partial}{\partial R}\left[R^{1/2}\frac{\partial}{\partial R}(\Delta\mu R^{1/2})\right]$$

Now look at the four terms in this equation carefully and compare them with Equation 86 above. Do you recognize that the first term on the left-hand side cancels with the first term on the right? (Remember, $\mu_0 = \nu_0\Sigma_0$!) In fact, we are left with

$$\frac{\partial(\Delta\Sigma)}{\partial t} = \frac{3}{R}\frac{\partial}{\partial R}\left[R^{1/2}\frac{\partial}{\partial R}(\Delta\mu R^{1/2})\right] \tag{87}$$

which is indeed the penultimate equation in FKR on page 116.

Although this is looking good already – we evidently have an equation involving just the perturbation we applied to the disc – something is still missing. For the equation involves two unknowns, $\Delta\Sigma$ and $\Delta\mu$. Can we get rid of one of them? Yes, if we restrict ourselves to *small* perturbations. We need to know by what amount $\Delta\mu$ the quantity μ changes locally, at a certain disc radius R, when the surface density Σ changes by an amount $\Delta\Sigma$. You certainly recognize that for small changes $\Delta\Sigma$ this is exactly what the derivative $\partial\mu/\partial\Sigma$ is meant to express (we use partial derivatives ∂ rather than 'd' because this has to be taken at a constant disc radius and at a fixed time). So from

$$\frac{\partial\mu}{\partial\Sigma} \approx \frac{\Delta\mu}{\Delta\Sigma}$$

we have $$\Delta\mu = \left(\frac{\partial\mu}{\partial\Sigma}\right)\Delta\Sigma \tag{88}$$

where the derivative $\partial\mu/\partial\Sigma$ refers to the local change of μ with Σ in a disc ring at radius R with surface density Σ_0. Now the trick is that (a) as long as the perturbation $\Delta\Sigma$ is sufficiently *small* Equation 88 is essentially an exact relation between $\Delta\Sigma$ and $\Delta\mu$, and that (b) as long as the perturbation is sufficiently *localized*, in the vicinity of R, $\partial\mu/\partial\Sigma$ does not depend on R either. It is a *constant*! This is why the technique we are going through here is called a **local linear stability analysis**.

By writing Equation 88 as

$$\Delta\Sigma = \Delta\mu \bigg/ \left(\frac{\partial\mu}{\partial\Sigma}\right) = \Delta\mu \times \left(\frac{\partial\mu}{\partial\Sigma}\right)^{-1}$$

the time derivative of $\Delta\Sigma$ becomes

$$\frac{\partial(\Delta\Sigma)}{\partial t} = \frac{\partial}{\partial t}\left[\left(\frac{\partial\mu}{\partial\Sigma}\right)^{-1}\Delta\mu\right] = \left(\frac{\partial\mu}{\partial\Sigma}\right)^{-1}\frac{\partial(\Delta\mu)}{\partial t}$$

Remember, $\partial\mu/\partial\Sigma$ is constant and is therefore not affected by the derivative. Inserting this into Equation 87 we obtain

$$\left(\frac{\partial\mu}{\partial\Sigma}\right)^{-1}\frac{\partial(\Delta\mu)}{\partial t} = \frac{3}{R}\frac{\partial}{\partial R}\left[R^{1/2}\frac{\partial}{\partial R}(\Delta\mu R^{1/2})\right]$$

Solving for $\partial(\Delta\mu)/\partial t$ this finally gives

$$\frac{\partial(\Delta\mu)}{\partial t} = \frac{\partial\mu}{\partial\Sigma}\frac{3}{R}\frac{\partial}{\partial R}\left[R^{1/2}\frac{\partial}{\partial R}(\Delta\mu R^{1/2})\right] \tag{89}$$

the last equation on page 116 of FKR, which was the starting point of this exercise! Time to take a deep breath!

Question 79

Describe in words how we arrived at a diffusion equation for the perturbed quantities $\Delta\Sigma$ and $\Delta\mu$. ■

At this point it is perhaps useful to reiterate the main properties of the quantities $\Delta\Sigma$, $\Delta\mu$ and $\partial\mu/\partial\Sigma$ which we have used so extensively in the above stability analysis. The perturbation $\Delta\Sigma(R, t)$ of the surface density describes the difference between the surface density in a steady-state disc and the actual surface density in the perturbed disc, for each radius R and as a function of time t. To facilitate the analysis we had to consider the behaviour of the quantity $\mu = \nu\Sigma$, which is also a function of R and t. In particular, we investigated the perturbation $\Delta\mu(R, t)$ which describes the difference between the value of μ in a steady-state disc and the actual value of μ in the perturbed disc, for each radius R and as a function of time t. In the *linear* regime, i.e. for small perturbations, $\Delta\mu$ is simply given by the product $\Delta\Sigma$ times the derivative $\partial\mu/\partial\Sigma$. Strictly speaking, this derivative is also a function of radius R (but not of time). The derivative $\partial\mu/\partial\Sigma$ measures the change of μ relative to the change of Σ, when two steady-state discs with slightly different surface density profiles are compared, for each radius R. In the *local* stability analysis we focus on a very small section of the disc with a very small radial extent, so that $\partial\mu/\partial\Sigma$ can be treated as constant over the corresponding range of radii.

Question 80

State the main assumptions for a local linear stability analysis. ■

So now you know how to derive Equation 89, and you also know what the solutions $\Delta\mu(R, t)$ of this equations look like (Figure 74). Their character depends on the sign of $\partial\mu/\partial\Sigma$. If this derivative is positive ($\partial\mu/\partial\Sigma > 0$) Equation 89 is a diffusion equation of the type we have already investigated in Section 4.1.3. Any perturbation $\Delta\mu(R, t)$ will decay away on the viscous time (left panel of Figure 74). In this case the disc is viscously stable. Conversely, if $\partial\mu/\partial\Sigma < 0$ a perturbation $\Delta\mu(R, t)$ will *grow* (right panel of Figure 74), and the disc is viscously unstable.

But how can you recognize if a given accretion disc is viscously unstable or not? How do we actually calculate the gradient $\partial\mu/\partial\Sigma$?

Question 81

Look up Equations 5.19 and 5.43 in FKR and explain how they relate to the quantity $\mu = \nu\Sigma$. ■

The answer to Question 81 shows that $\mu \propto T^4$ and also $\mu \propto \dot{M}$. Hence the criterion for viscous stability, $\partial\mu/\partial\Sigma > 0$ (Equation 85) can also be written as

$$\text{viscously stable if} \quad \frac{\partial T(R)}{\partial\Sigma} > 0 \quad \text{or} \quad \frac{\partial \dot{M}(R)}{\partial\Sigma} > 0 \qquad (90)$$

Let us be clear about the meaning of these derivatives. First, they have to be taken at a fixed disc annulus with radius R. Second, they check how the equilibrium values of T or $\dot{M}$ change when Σ is varied. In order to work out these derivatives we have to compare, at a fixed radius R, steady-state disc models with different mass accretion rates.

The second form of the stability criterion (Equation 90) is intuitively clear. If we have $\partial \dot{M} / \partial \Sigma < 0$, then small changes in $\dot{M}$ and Σ are related as $\Delta \dot{M} \propto -\Delta \Sigma$. A perturbation that locally increases the surface density ($\Delta\Sigma > 0$) causes a corresponding decrease of the local mass accretion rate ($\Delta \dot{M} \propto -\Delta \Sigma < 0$). But the mass supply rate from adjacent disc regions has still the original higher value. Hence mass accumulates locally, and Σ increases further. Even if the stability analysis we have just forced on you is still a mystery to you, try to remember this last point.

You will investigate the far-reaching implication of a disc that violates the stability criterion in the multimedia Activity 48 very shortly.

We conclude this subsection with the last piece of vital information you need in order to understand the mysteries of accretion disc outbursts.

Activity 47

Thermal-viscous instability

Have a last mini-dose of FKR Section 5.8. On page 118, read the paragraph which contains FKR Equation 5.79 ('We must now …' to '… thermal instability').

FRANK KING RAINE

Activity 47 FKR Section 5.8

Keywords: **thermal-viscous instability** ■

Question 82

Derive Equation 5 in the set FKR Equation 5.74 from the statement that heating and cooling are balanced. (*Hint*: Recall the definition of Q^+ and Q^-. Then equate Q^+ and Q^-.) ■

A disc which is subject to a viscous instability will not be able to remain in thermal equilibrium. Rather, a thermal instability will occur as well, allowing the disc to reach another viscous equilibrium, i.e. a state which is viscously stable.

This can also be expressed as follows

$$\frac{\partial T}{\partial t} \propto Q^+ - Q^- \qquad \text{(FKR 5.79) (91)}$$

This could be called a 'phenomenological equation' as it is not the result of a strict derivation. It purely expresses the expectation that the disc temperature must rise ($\partial T/\partial t > 0$) if the heating dominates the cooling ($Q^+ > Q^-$, i.e. $Q^+ - Q^- > 0$), and conversely, the temperature must drop ($\partial T/\partial t < 0$) if cooling dominates the heating.

Example 26

Why does a small viscous perturbation exaggerate a mismatch of Q^+ and Q^- if the disc is viscously unstable? (Note that $\nu \propto T$.)

Solution

Assume that the perturbation is such that the surface density increases a bit, by $\Delta\Sigma > 0$. Then as a consequence the disc temperature must drop somewhat, for viscous instability means that $\partial T/\partial\Sigma < 0$ (Equation 90). As

$$\frac{\Delta T}{\Delta \Sigma} \approx \frac{\partial T}{\partial \Sigma}$$

we have $\Delta T \approx \dfrac{\partial T}{\partial \Sigma} \Delta \Sigma < 0$

This drop in temperature implies a drop in the cooling rate, $Q^- \propto T^4$. At the same time the increased surface density implies that the heating rate $Q^+ \propto \mu = \nu \Sigma$ usually increases. Q^+ might also decrease: because $\nu \propto T$, the reduced temperature implies a reduced viscosity. But the decrease of Q^- is in any case more pronounced. So cooling dominates heating, and the temperature will drop even further. ■

[handwritten annotation:] Errata
Hence the reaction of the temperature to the change in surface density exaggerates the mismatch between heating + cooling.

We can summarize this also in a slightly different way: The viscous instability as we defined it does not occur, since the assumption of thermal equilibrium that we have made to define it breaks down. ~~Instead, a thermal instability occurs, thus allowing the disc to evolve on a thermal timescale into another, viscously stable state.~~

You are now well prepared for the main interactive multimedia tutorial (Activity 48), the moment you have been waiting for.

Activity 48	(4 hours)

[handwritten annotation:] Activity 48
Accretion disc outburst
(Multimedia tutorial)

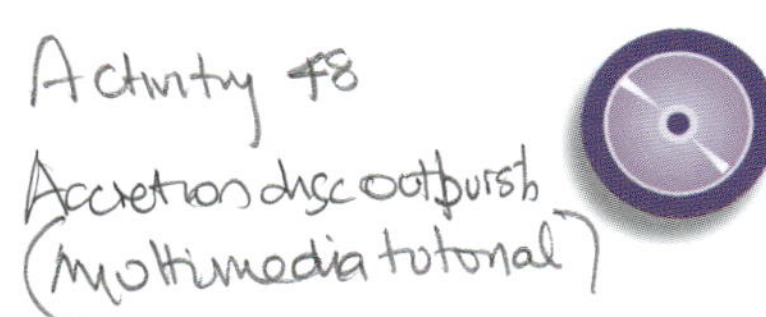

Accretion disc outbursts

You should now study the main multimedia tutorial package for this block.

The agenda of the tutorial is:

- Review the outburst light curves of systems known to have accretion discs (dwarf novae, soft X-ray transients).

- Revise and plot radial surface density profiles $\Sigma(R)$ of steady-state discs with radiative and convective energy transport. (You met these earlier in Activity 29.) Zoom-in on one disc annulus and watch how the surface density changes when the mass accretion rate is varied. Discover that there is a limited regime of accretion rates where the surface density is triple-valued. Identify hydrogen recombination as the physical cause for this.

- Construct the S-curve, i.e. plot \dot{M} versus Σ. Discuss the viscous stability of the three branches of the S-curve. Recall viscous diffusion and how a perturbation could lead to either viscous decay (stability) or fragmentation (instability).

- Discuss the local reaction of the disc to perturbations from the equilibrium S-curve, recalling the hierarchy of timescales.

- Understand the local limit cycle using the S-curve (animation).

- Watch the global outburst cycle via animated $\Sigma(R)$-profiles; relate different phases of the cycle to the corresponding section of the light curves.

- Stress that outburst timescales relate to viscous timescales and therefore constrain viscosity.

Keywords: none ■

The name game: part 8

Skim-read Section 8 and make a list of all symbols that denote *new* variables and *new* quantities, and note down what they mean. Flag up those symbols that have (unfortunately) multiple uses.

Once you have compiled your list and explained the meaning of all entries, compare it to the one given in Appendix A5 at the end of the Study Guide.

8.3 Summary of Section 8

1 Hydrostatic equilibrium is established on the vertical sound crossing time $t_z = H/c_S$. The timescale t_z is of order the dynamical time $t_\phi = \Omega_K^{-1}$. Thermal equilibrium is established on the thermal timescale t_{th}. In thermal equilibrium the local cooling by radiation and convection is exactly balanced by the local viscous heating.

2 There is a hierarchy of timescales,

$$t_\phi \approx t_z \approx \alpha t_{th} \approx \alpha \left(\frac{H}{R}\right)^2 t_{visc}$$

i.e. for $\alpha < 1$ the dynamical timescale is shorter than the thermal timescale, while both are much shorter than the viscous timescale t_{visc}. The viscous timescale is the characteristic timescale of the radial drift of matter in the disc.

3 In a steady-state accretion disc the local heating rate Q^+ (by viscous dissipation) just equals the local cooling rate Q^- (by energy transport in the vertical direction).

4 The local disc heating rate per unit volume (measured, for example, in units of erg s^{-1} cm^{-3}) is approximately given by

$$Q^+ \approx \frac{D(R)}{H}$$

where $D(R)$ is the local dissipation rate per unit surface area, and H the disc scale height.

5 If the vertical energy transport is by radiative diffusion the local cooling rate per unit volume (measured in units of erg s^{-1} cm^{-3}) is approximately given by

$$Q^- \approx \frac{4\sigma T_c^4}{3\kappa\rho H^2}$$

6 A steady-state disc in thermal equilibrium is viscously stable if an accidental density perturbation decays away on the viscous time. If the disc is viscously unstable the perturbation grows and causes the disc to break up into detached rings.

7 The criterion for viscous stability can be written as

$$\frac{\partial T(R)}{\partial \Sigma} > 0 \quad \text{or} \quad \frac{\partial \dot{M}(R)}{\partial \Sigma} > 0$$

8 A disc is thermally unstable if an (accidental) temperature perturbation keeps growing on a thermal timescale just by itself. This can be expressed as

$$\frac{\mathrm{d}(Q^+ - Q^-)}{\mathrm{d}T_c} > 0$$

9 A disc subject to a viscous instability will also develop a thermal instability.

10 Local linear stability analysis considers the reaction of systems to localized, small perturbations.

11 Geometrically thin, optically thick, Keplerian steady-state accretion discs are viscously unstable in the region where hydrogen is partially ionized, i.e. at temperatures around $T_H \sim$ 6000–8000 K.

12 Dwarf nova and soft X-ray transient outbursts can be understood as the result of this thermal-viscous instability.

13 The disc undergoes a limit-cycle evolution if it is subject to this instability. The local instability launches heating and cooling waves that propagate through the disc, making the disc subject to a global instability. The disc alternates between a hot, bright state with a large mass accretion rate and a cool, dim state with small accretion rate.

14 In the hot state the accretion disc is in a quasi-steady state. The disc slowly loses mass as the mass accretion rate through the disc is larger than the mass supply rate from the donor star. The hot state duration is a fraction of the viscous time in the hot state.

15 In the cool state the disc accumulates mass and slowly evolves on the viscous time of the cold state. The disc may be optically thin, so that the Shakura–Sunyaev description is not an accurate model.

16 The viscosity in the cool state must be much smaller than in the hot state ($\alpha_{\rm cold} \ll \alpha_{\rm hot}$).

9 CONSOLIDATION AND REVISION

9.1 The world is not enough

As you must be acutely aware, we have simply run out of time to continue further on our tour of the events taking place in accreting systems. Should you feel like browsing the uncharted territory of FKR (*but only after you have completed the course*!) you are likely to find a wealth of hints on other exciting phenomena, all at the forefront of astrophysical research.

Areas of intense research are the origin of viscosity and the physical nature of accretion flows in regimes different than the one we discussed in detail. Discs may, for instance not be geometrically thin. Or they may extend down to the vicinity of a black hole event horizon, where general relativity has a say in how matter and radiation behave. In discs which become very hot, radiation pressure plays an important role, giving rise to a new type of instability. Other problems occur when the disc is not simply optically thick and in local thermodynamic equilibrium. Thermal energy might be trapped in the accreting plasma and carried away from the point where it has been generated, simply due to the radial drift of matter. This is called advection. Yet other effects occur if the disc is self-irradiated.

With increasingly detailed Doppler tomograms of accreting systems it is feasible to probe more complicated structures superimposed on the discs, such as spiral shocks and warps. Research on systems with extreme mass ratios reveal fascinating results. The disc is seen to become slightly eccentric and starts to precess, i.e. rotate slowly in the binary frame. Theoretical models for non-axisymmetric discs involve the study of two- or three-dimensional hydrodynamic flows and are increasingly complex. The detailed modelling of astrophysical fluids continues to challenge the most powerful computers.

9.2 Looking back

At this point, however, you should sit back and revise and consolidate what you have learnt in this block. Clearly, there have been two main themes, one on binary stars and their evolution, the other on accretion discs and their appearance. You will have the opportunity to consolidate your knowledge in each of these areas in the following four activities.

Activity 49 and Activity 50 will guide you to two short review articles in the corresponding fields. These articles are not written for a popular audience, but aimed at astrophysicists who wish to update their knowledge in fields where they are not necessarily experts. Therefore you should *not* expect that the articles are *easy* to read. What is more, they will contain some new concepts and ideas, and may use a slightly different notation from what you are used to. You are *not* expected to remember – or in fact not really fully understand – these new concepts. The review articles should give you a slightly different perspective on the issues that we have discussed in the block. When you read the reviews, compare them with the material in the corresponding sections of the Study Guide.

Activity 51 and Activity 52 will direct you to websites that contain material relevant to this block.

As a first step in your revision week you may wish to read the summaries of the preceding eight main sections. Reflect on each item in the summary lists. Repeat the statements in your own words, and expand on them as if you had to explain the underlying concepts, ideas and implications to somebody who has a general background in astrophysics, but who is not an expert in interacting binary stars. It is quite likely that you will have difficulty recalling the context and importance of quite a few of these summary items. There is no reason to be afraid if this is the case. Rather, go back to the relevant individual subsections and find the corresponding discussion in the Study Guide. If necessary, read also the corresponding paragraphs in FKR again.

Activity 49 **(90 minutes)**

A review of binary evolution

Connect to the S381 home page, select Block 3, Activity 49 and follow the instructions on screen.

Keywords: none ■

Activity 50 **(90 minutes)**

A review of accretion discs

Connect to the S381 home page, select Block 3, Activity 50 and follow the instructions on screen.

Keywords: none ■

At this point you may wish to study the main multimedia tutorial (Activity 48) on accretion disc outbursts again.

Finally, there are two more websites for you to look at.

Activity 51 **(30 minutes)**

Binary stars on the Web

Connect to the S381 home page, select Block 3, Activity 51 and follow the instructions on screen.

Keywords: none ■

Activity 52 **(30 minutes)**

Accretion on the Web

Connect to the S381 home page, select Block 3, Activity 52 and follow the instructions on screen.

Keywords: none ■

ACHIEVEMENTS

Now that you have completed Block 3, you should be able to:

A1 Explain the meaning of the all the newly defined (emboldened) terms introduced in this block, and understand these when they are used in scientific articles not specifically written for a student audience.

A2 Carry out activities using the computer, in particular using the World Wide Web, the Image Archive, and spreadsheet calculations.

A3 Manipulate numbers and symbols that occur in equations describing the physics of close binary stars and accretion.

A4 Appreciate the power of the mathematical language in describing physical systems and phenomena.

A5 Express in physical terms the meaning of equations describing physical systems.

A6 Perform simple manipulations on ordinary and partial differential equations.

A7 Carry out order of magnitude estimates to obtain characteristic timescales and dimensions of physical systems described by differential equations.

A8 Appreciate what setting up a physical model involves, and what it does not involve.

A9 Understand the interaction between a physical model and experiment.

A10 Recognize why binary stars are common, and why higher-order multiples occur only as hierarchical binaries.

A11 Understand the opportunities to probe stellar parameters offered by the presence of a stellar companion.

A12 Explain why mass transfer must occur in close binary systems.

A13 Explain why mass transfer to a compact star leads to the formation of a hot accretion disc that emits high-energy radiation.

A14 Appreciate that accretion discs are like machines that extract gravitational potential energy and angular momentum from plasma, and that accretion is the most efficient energy source in the Universe.

A15 Describe the physical nature of, and the characteristics of, the different types of interacting binary stars, in particular Algols, cataclysmic variables, X-ray binaries, and the subtypes of the latter two groups (or classes).

A16 Explain the Roche model.

A17 Appreciate the different modes of mass transfer, and the different mechanisms that drive mass transfer.

A18 Calculate the steady-state mass transfer rate, and explain what unstable mass transfer means.

A19 Explain the first step in the formation of an accretion disc from Roche-lobe overflow.

A20 Appreciate the importance of viscosity for accretion.

A21 Define and use vertically integrated variables to describe the radial structure of accretion discs.

A22 Explain viscous torque and viscous dissipation as a result of viscosity.

A23 Explain what is meant by the α-viscosity.

A24 Apply the conservation laws of mass, angular momentum and energy in the context of accretion flows.

A25 Demonstrate an understanding of the process of diffusion, and appreciate that it can be described by the diffusion equation.

A26 Understand viscous diffusion, the viscous timescale and radial drift.

A27 Explain the concept of steady-state accretion, and appreciate how it simplifies the description of accretion flows.

A28 Define what is meant by the terms geometrically thin, optically thick, and steady state in the context of accretion discs.

A29 Calculate the accretion luminosity, the accretion disc luminosity, and the boundary layer luminosity.

A30 Appreciate that accretion discs around white dwarfs emit predominantly in the ultraviolet, discs in neutron star or black hole systems are dominant in the X-ray region of the electromagnetic spectrum.

A31 Characterize the continuum spectrum of accretion discs.

A32 Appreciate under what conditions an accretion disc is geometrically thin.

A33 Explain the difference between energy transport by radiative diffusion and convection.

A34 Explain what is meant by a Shakura–Sunyaev disc model, and appreciate the typical dimensions of accretion discs in cataclysmic variables.

A35 Appreciate that accretion can change the spin rate of the accreting star.

A36 Describe observational signatures of accretion discs. This includes double-peaked emission lines and a characteristic shape of the light curve.

A37 Understand and describe the eclipse mapping technique.

A38 Explain the broad principles on which Doppler tomography is based.

A39 Appreciate the differences between accretion discs in cataclysmic variables and in X-ray binaries.

A40 Appreciate the concept that magnetic fields are frozen in to a perfectly conducting plasma.

A41 Describe the magnetic field structure of a magnetic dipole.

A42 Calculate magnetic pressure and ram pressure, and define Alfvén radius or magnetospheric radius.

A43 Appreciate observable effects of magnetically controlled accretion.

A44 Explain the hierarchy of timescales in Keplerian accretion discs.

A45 Explain the concept of thermal stability and viscous stability of an accretion disc.

A46 Appreciate what is meant by local linear stability analysis.

A47 Recognize the importance of hydrogen recombination for the viscous instability of accretion discs.

A48 Explain the basic principles of the disc instability model for dwarf nova and soft X-ray transient outbursts.

ANSWERS TO QUESTIONS

Question 1

If an object shines at the Eddington luminosity, the outward radiation pressure force just balances the inward gravitational force. The Eddington luminosity is the maximum luminosity of a gravitationally bound object. The accretion luminosity is the rate at which energy is liberated when an object accretes matter.

Question 2

(a) The Eddington luminosity is given by

$$L_{\text{edd}} = 1.3 \times 10^{38} \left(\frac{M}{\text{M}_\odot} \right) \text{erg s}^{-1}$$

For the Sun we have $L_{\text{edd}} = 1.3 \times 10^{38}$ erg s^{-1} = $3.4 \times 10^4 \text{L}_\odot$. (The solar luminosity is $1\text{L}_\odot = 3.83 \times 10^{33}$ erg s^{-1}.) A neutron star with mass $1\text{M}_\odot$ has the same Eddington luminosity.

(b) The Eddington accretion rate ($\dot{M}_{\text{edd}}$) is obtained by equating $L_{\text{edd}} = L_{\text{acc}}$. Hence with

$$L_{\text{acc}} = \frac{GM\dot{M}}{R_*}$$

we have for the Sun, in cgs units (note that erg = dyne cm)

$$\dot{M}_{\text{edd}} = \frac{R_* L_{\text{edd}}}{GM} = \frac{6.96 \times 10^{10} \text{ cm} \times 1.3 \times 10^{38} \text{ erg s}^{-1}}{6.673 \times 10^{-8} \text{ dyne cm}^2 \text{ g}^{-2} \times 1.99 \times 10^{33} \text{ g}}$$

$$= 6.8 \times 10^{22} \text{ g s}^{-1}$$

For the neutron star with radius 10 km we have

$$\dot{M}_{\text{edd}} = (\dot{M}_{\text{edd}})_{\text{Sun}} \times \frac{1 \times 10 \text{ km}}{1\text{R}_\odot} = 6.8 \times 10^{22} \text{ g s}^{-1} \times \frac{1 \times 10^6 \text{ cm}}{6.96 \times 10^{10} \text{ cm}}$$

$$= 9.8 \times 10^{17} \text{ g s}^{-1}$$

As we have

$$1\text{M}_\odot \text{ yr}^{-1} = \frac{1.99 \times 10^{33} \text{ g}}{365.25 \times 86\,400 \text{ s}} = 6.3 \times 10^{25} \text{ g s}^{-1} \tag{92}$$

the Eddington rate for the Sun is a huge $10^{-3}\text{M}_\odot \text{ yr}^{-1}$, while for the neutron star it is $1.6 \times 10^{-8}\text{M}_\odot \text{ yr}^{-1}$.

Question 3

The accretion efficiency η is defined by $L_{\text{acc}} = \eta \dot{M} c^2$.

Equating this to Equation 2 and solving for η gives

$$\eta = \frac{GM}{R_* c^2}$$

For a neutron star with mass $1\mathrm{M}_\odot = 1.99 \times 10^{33}$ g and radius 10 km = 10^6 cm this is

$$\eta = \frac{6.673 \times 10^{-8}\ \mathrm{dyne\,cm^2\,g^{-2}} \times 1.99 \times 10^{33}\ \mathrm{g}}{1 \times 10^6\ \mathrm{cm} \times (2.998 \times 10^{10}\ \mathrm{cm\,s^{-1}})^2} \approx 0.15$$

Question 4

(a) The mass defect $\Delta m = 4.40 \times 10^{-29}$ kg involved in the fusion of four protons into one helium nucleus translates into an energy gain of $\Delta E = \Delta mc^2$ per four protons. The energy input is the mass-energy of the four protons, $4m_\mathrm{p}c^2$, so the efficiency = gain/input is

$$\eta = \frac{\Delta mc^2}{4m_\mathrm{p}c^2} = \frac{4.40 \times 10^{-29}\ \mathrm{kg}}{4 \times 1.673 \times 10^{-27}\ \mathrm{kg}} \approx 0.0066$$

(b) Hence the efficiency of hydrogen burning, the most common nuclear fusion reaction in the Universe, is only $\eta = 0.007$. In other words, if one gram of hydrogen accretes onto a compact star (neutron star or black hole) it liberates 20 times more energy (in the form of heat and radiation, say) than if this gram of hydrogen undergoes nuclear fusion into helium.

There is no other process in the Universe which could sustain the conversion of such a large fraction of mass-energy ($\gtrsim$10%) into energy for a long time for a macroscopic amount of mass. (You may have heard about processes with 100% efficiency such as the annihilation of electron–positron pairs. But these involve antimatter which is not abundant in the known Universe.)

Question 5

We have 1 eV = 1.602×10^{-12} erg, and $T = E_\mathrm{ph}/k$. So for E_ph = 1 eV the temperature is

$$T = \frac{1\,\mathrm{eV}}{1.381 \times 10^{-16}\ \mathrm{erg\,K^{-1}}} = \frac{1.602 \times 10^{-12}\ \mathrm{erg}}{1.381 \times 10^{-16}\ \mathrm{erg\,K^{-1}}} = 1.160 \times 10^4\ \mathrm{K}$$

which is of order 10^4 K.

Question 6

The virial temperature is essentially the temperature the accreted material would reach if its gravitational potential energy were turned entirely into thermal energy. The blackbody temperature is the temperature the source would have if it were to radiate the accretion luminosity as a blackbody spectrum. The blackbody temperature is usually lower than the virial temperature. The significance of these temperatures is that they give an indication of the expected radiation temperature, i.e. the mean energy of the photons the accreting gas emits. If the gas is opaque (optically thick), the energy of both photons and particles in the gas is thermalized, i.e. distributed over a wide range of energies. In this case the radiation temperature is the blackbody temperature. If the gas is transparent (optically thin), then the photon energy distribution is much narrower. The corresponding radiation temperature is the virial temperature.

Question 7

The outburst amplitude of SS Cyg is a few magnitudes, say $\Delta m_V = 4$ mag. The magnitude scale is logarithmic,

$$\Delta m_V = m_1 - m_2 = 2.5 \log\left(\frac{L_2}{L_1}\right)$$

(see Block 1, Equation 51). So the luminosity amplitude $A = L_2/L_1$ is

$$A = \frac{L_2}{L_1} = 10^{\Delta m_V / 2.5} = 10^{4/2.5} = 10^{1.6} = 40$$

In classical novae, the outburst amplitude more typically is $\Delta m_V = 8$ mag or even 10 mag (see, for example, the one shown in Figure 12). The corresponding increase in the luminosity is $10^{10/2.5} = 10^4$!

Question 8

There are images of two nova remnants in the Image Archive: Nova Cygni 1992 and T Pyxidis. At low resolution (early HST image of N Cyg, and ground-based image of T Pyx) both nebulae appear fairly smooth and almost spherical. At higher resolution more structure becomes apparent. T Pyx is very inhomogeneous (clumpy); N Cyg develops a pronounced shell structure.

Supernova remnants are generally more irregular and larger (both in terms of their actual and angular size) than nova shells. Planetary nebulae (PNe) are usually very symmetric, but the PNe morphology is very diverse. Some PNe resemble nova shells.

Question 9

The nova shell is seen to expand. In the 7 months between the two observations the mean radius has increased by a tiny but significant 0.037″. From radial velocity measurements we know that the true expansion velocity is about 970 km s^{-1}. The true radius increase of the nova shell in 7 months therefore is

$$\Delta r = v_{\rm exp}\Delta t = 970 \times 10^5 \text{ cm s}^{-1} \times 7 \text{ months} \times 30 \text{ day month}^{-1} \times 86\,400 \text{ s day}^{-1}$$
$$= 1.76 \times 10^{15} \text{ cm}$$

If the distance to the nova is D, then using the small-angle approximation the angle $\delta = \Delta r/D$ is just the measured 0.037″. So, converting the angle into radians,

$$D = \frac{\Delta r}{\delta} = \frac{1.76 \times 10^{15} \text{ cm}}{0.037 \times (1/3600)(\pi/180)} = \frac{1.76 \times 10^{15} \times 3600 \times 180}{0.037 \times \pi} \text{ cm}$$
$$= 9.8 \times 10^{21} \text{ cm}$$

The distance, therefore, is 3.2 kpc (1 pc = 3.086×10^{18} cm).

Question 10

(a) The observations have been made in the radio band, at wavelength 3.6 cm.

(b) The outburst occurred on 18 March 1994, the observations cover the period up to 16 April 1994. The last observation has been made 28 days after the outburst.

(c) The angular expansion is $\delta = 0.5''$ in each jet.

(d) The implied linear expansion is $\Delta l = D \tan \delta$. The distance is $D \approx 12.5\,\text{kpc}$. Hence, naively, the implied velocity would simply be $v = \Delta l/(28\ \text{days})$, i.e.

$$v = \frac{(\tan 0.5'') \times 12.5 \times 10^3 \times 3.086 \times 10^{18}\,\text{cm}}{28 \times 86\,400\,\text{s}} = 3.87 \times 10^{10}\,\text{cm}\,\text{s}^{-1}$$

This is 1.29 times the speed of light in vacuum.

Question 11

Dwarf novae are cataclysmic variables (systems with a white dwarf accretor), that alternate semiregularly between outburst and quiescence on timescales of weeks. The outburst mechanism is an instability of the accretion disc. Classical nova outbursts occur also in CVs. They constitute the explosive burning of hydrogen to helium on the surface of the white dwarf. Classical novae have much larger outburst amplitudes than dwarf novae, and no classical nova system has been observed to go into outburst more than once. If nova outbursts do occur in the same system, the recurrence time must be very long, certainly longer than 10^3 years. (Theoretical calculations predict that the recurrence time is 10^4 to 10^5 years.)

If the accreting star is a neutron star or black hole, the system appears as an X-ray binary. X-ray novae are the X-ray binary analogue of dwarf novae, i.e. the outbursts are due to an instability in the accretion disc around the neutron star or black hole. Type I X-ray bursts are the neutron star X-ray binary analogue of classical novae, i.e. the outbursts are due to a thermonuclear runaway on the surface of a neutron star. But compared to the white dwarf case the timescales are inverted, X-ray bursts occur very frequently (with recurrence times hours to days), while X-ray nova outbursts are usually observed only once per system (recurrence time decades or longer) – see Table 1.

A type Ia supernova is the disruption of a white dwarf that is driven over the Chandrasekhar mass. This is likely to occur due to mass accretion in a binary system. A type Ia supernova cannot recur in a given system.

Question 12

The centrifugal force on a body with mass m at distance r from the rotational axis is $F_c = m\omega^2 r$, while the gravitational force is $F_{GR} = mg$ ($g = 9.81\,\text{m}\,\text{s}^{-2}$). The angular velocity is $\omega = 2\pi/(86\,400\,\text{s})$; at the equator we have $r = 6.4 \times 10^6\,\text{m}$. Hence at the equator the force ratio is

$$\frac{F_c}{F_{GR}} = \frac{\omega^2 r}{g} = \frac{(2\pi)^2 \times 6.4 \times 10^6}{86\,400^2 \times 9.81} = 3.5 \times 10^{-3}$$

The North and South Pole are on the rotational axis of Earth, i.e. $r = 0$, $F_c = 0$. The ratio F_c/F_{GR} is 0.

(Note that the force ratio is independent of the mass of the body.)

Question 13

FKR Equation 4.5 reads

$$\Phi_R(\boldsymbol{r}) = -\frac{GM_1}{|\boldsymbol{r} - \boldsymbol{r}_1|} - \frac{GM_2}{|\boldsymbol{r} - \boldsymbol{r}_2|} - \frac{1}{2}(\boldsymbol{\omega} \times \boldsymbol{r})^2$$

This is the Roche potential at a point – we call it the reference point – fixed in the co-rotating frame (binary frame). The position of this point of reference is described by the position vector $\boldsymbol{r}$. The force on a test body with mass m at rest in the binary frame can be calculated as the gradient of the Roche potential. To be precise, the force is given by $-m\nabla\Phi_{\rm R}$.

On the right-hand side, the first term is the gravitational potential of the primary star. The denominator is the magnitude of the vector pointing from the primary to the point of reference. The second term is the corresponding gravitational potential of the secondary. The third term describes the effect of the centrifugal force. The quantity $(\boldsymbol{\omega}\times\boldsymbol{r})^2$ is the scalar product of the vector $\boldsymbol{\omega}\times\boldsymbol{r}$ with itself. The vector $\boldsymbol{\omega}\times\boldsymbol{r}$ has the magnitude $\omega r_\perp$ where $r_\perp$ is the distance of the point of reference from the rotational axis. The vector $\boldsymbol{\omega}$ is parallel to the rotational axis and has magnitude ω, the orbital angular speed.

Question 14

The main assumptions are (a) the orbit is a circle, (b) the two components are point masses, (c) the stars rotate synchronously with the orbit.

Assumption (b) is a good approximation even for the Roche-lobe filling component if this is sufficiently centrally condensed – as most stars are. Assumption (c) is justified for Roche-lobe filling stars as these are efficiently tidally locked to the orbit. (See also the box on 'Magnetic braking' in Section 2.2.)

Question 15

A saddle point of a potential is a point where the spatial gradient of the potential Φ vanishes, such that the potential is a maximum in one direction, e.g. along the x-axis, but a minimum in a direction perpendicular to the former, i.e. along the y-axis. Mathematically,

$$\frac{\partial\Phi}{\partial x} = \frac{\partial\Phi}{\partial y} = 0$$

and the second partial derivative $\partial^2\Phi/\partial x^2$ is negative, while $\partial^2\Phi/\partial y^2$ is positive.

Question 16

There are five points in space – the Lagrangian points L_1, L_2, L_3, L_4 and L_5 – where the Roche potential reaches a local maximum or minimum. The points L_1 to L_3 all lie on the line that connects the stellar centres, and the Roche potential along this line reaches a local maximum in each of these points. L_1 is the familiar inner Lagrangian point, while L_2 and L_3 are so-called outer Lagrangian points. The potential at L_2 and L_3 is larger than at L_1.

The outer Lagrangian points L_4 and L_5 are also in the orbital plane and form an equilateral triangle with the two binary components. At L_4 and L_5 the Roche potential reaches a maximum.

Question 17

The conversion rate is $1{\rm M}_\odot\,{\rm yr}^{-1} = 6.3\times10^{25}\,{\rm g\,s}^{-1}$ (see Equation 92, in the answer to Question 2). So the case A transfer rates are $6\times10^{15}\,{\rm g\,s}^{-1}$ for the $1{\rm M}_\odot$ star and $4\times10^{18}\,{\rm g\,s}^{-1}$ for the $5{\rm M}_\odot$ star. The corresponding case B rates are $6\times10^{18}\,{\rm g\,s}^{-1}$ for the $1{\rm M}_\odot$ star and $4\times10^{21}\,{\rm g\,s}^{-1}$ for the $5{\rm M}_\odot$ star.

Question 18

(a) Kepler's law is

$$\frac{a^3}{P^2} = \frac{GM}{4\pi^2}$$

while Paczyński's approximation for the Roche lobe radius is

$$\frac{R_2}{a} = 0.462\left(\frac{M_2}{M}\right)^{1/3}$$

We solve Kepler's law for the period,

$$P^2 = \frac{a^3 4\pi^2}{GM}$$

and multiply both the numerator and denominator on the right-hand side by $(R_2/a)^3$. Hence

$$P^2 = \frac{a^3 4\pi^2}{GM}\frac{(R_2/a)^3}{(R_2/a)^3} = \frac{R_2^3 4\pi^2}{GM}\frac{1}{(R_2/a)^3}$$

Inserting Paczyński's approximation for R_2/a gives

$$P^2 = \frac{R_2^3 4\pi^2}{GM}\frac{M}{0.462^3 M_2} = \frac{4\pi^2}{0.462^3 G}\frac{R_2^3}{M_2}$$

Taking the square root and noting that

$$\bar{\rho} = \frac{M_2}{(4\pi/3)\times R_2^3}$$

(the stellar radius equals the Roche lobe radius R_2) we have

$$P = \left(\frac{4\pi^2}{0.462^3 G}\right)^{1/2}\left(\frac{4\pi}{3}\bar{\rho}\right)^{-1/2}$$

With $G = 6.673 \times 10^{-8}$ dyne cm^2 g^{-2} we have

$$P = \frac{P}{1\,\text{h}}3600\,\text{s} = P_{\text{hr}} \times 3600\,\text{s}$$

$$= \left(\frac{3\pi}{0.462^3 \times 6.673 \times 10^{-8}\,\text{dyne}\,\text{cm}^2\,\text{g}^{-2}}\right)^{1/2}\sqrt{\frac{1}{(\bar{\rho}/\text{g}\,\text{cm}^{-3})\,\text{g}\,\text{cm}^{-3}}}$$

And hence

$$P_{\text{hr}} \cong \frac{10.5}{\sqrt{\bar{\rho}/\text{g}\,\text{cm}^{-3}}}$$

(Note that 1 dyne = 1 g cm s^{-2}.) This is Equation 15.

(b) The free-fall time (see Section 1.1.3 of Block 2).

(c) Equation 15 has been derived using Paczyński's approximation for R_2/a. Hence Equation 15 is valid only in the range of mass ratios q where this approximation is good. FKR says that this is the case for $0.1 \lesssim q \lesssim 0.8$, but the approximation is in fact acceptable for any q less than unity.

Question 19

The binary pulsar was discovered in 1974 with the Arecibo 300 m radio telescope. The observations were made in the radioband.

Question 20

We use

$$\frac{-\dot{M}_2}{M_2} = \frac{-\dot{J}/J}{4/3 - M_2/M_1}$$

for the mass transfer rate (FKR Equation 4.17), and

$$\frac{\dot{J}_{GR}}{J} = -1.27 \times 10^{-8} \frac{m_1 m_2}{(m_1 + m_2)^{1/3}} P_{hr}^{-8/3}\ \text{yr}^{-1}$$

for the angular momentum loss rate. We have $m_1 = 1$, and $m_2 = 0.11P_{hr} = 0.22$.

Hence $$\frac{\dot{J}_{GR}}{J} = -4.12 \times 10^{-10}\ \text{yr}^{-1}$$

and so $$\frac{-\dot{M}_2}{0.22\text{M}_\odot} = \frac{4.12 \times 10^{-10}\ \text{yr}^{-1}}{4/3 - 0.22/1}$$

This gives $|\dot{M}_2| = 8.14 \times 10^{-11}\text{M}_\odot\ \text{yr}^{-1}$ or $|\dot{M}_2| = 5.13 \times 10^{15}\ \text{g s}^{-1}$.

Question 21

(a) The sound speed is given by

$$c_s \cong 10\left(\frac{T}{10^4\ \text{K}}\right)^{1/2}\ \text{km s}^{-1}$$

From Figure 49 in Block 1 we estimate that the surface temperature of a $1\text{M}_\odot$ star is roughly 5000 K, while that of a $0.5\text{M}_\odot$ star is 3500 K. (There is no need to worry if you had slightly different estimates; the T-axis of the figure is in log units, so it is difficult to interpolate between the scale divisions by eye.) With these values for T one obtains $c_s \approx 7\ \text{km s}^{-1}$ for the $1\text{M}_\odot$ star, and $\approx 6\ \text{km s}^{-1}$ for the $0.5\text{M}_\odot$ star.

(b) The orbital speed of a test body with mass m on a circular orbit just above the photosphere of a star with radius R and mass M can be obtained from the balance of gravitational and centrifugal forces:

$$\frac{GMm}{R^2} = m\frac{v^2}{R}$$

so that $v = \sqrt{GM/R}$. For $M = 1\text{M}_\odot$, $R = 1\text{R}_\odot$ this gives

$$v = \sqrt{\frac{6.673 \times 10^{-8} \times 1.99 \times 10^{33}}{6.96 \times 10^{10}}}\ \text{cm s}^{-1} = 4.4 \times 10^7\ \text{cm s}^{-1} = 440\ \text{km s}^{-1}$$

while for $M = 0.5\text{M}_\odot$, $R = 0.5\text{R}_\odot$ we find the same value. So the orbital speed is almost 100 times faster than the sound speed, i.e. the orbital motion is highly supersonic.

Question 22

The circularization radius r_{circ} is the radius of a fictitious circular orbit centred on the accreting star, defined by the following property. Material orbiting the accreting star on this orbit with Keplerian angular speed ω has the same specific angular momentum (angular momentum per unit mass) $r_{\text{circ}}\omega^2$ with respect to the accreting star as material that is just about to leave the donor star through the inner Lagrangian point towards the accreting star.

Question 23

A rather famous differentially rotating body is the Sun. This can be seen when groups of sunspots move across the disc of the Sun. Sunspots at higher latitudes lag behind sunspots that are closer to the equatorial region. Hence the angular velocity in equatorial regions is larger than in polar regions.

An indirect example of differential rotation can be seen when sprinters in separate lanes follow the curve of a stadium (e.g. in a 400 m heat). Even if the athletes in the inner and outer lanes run at the same speed, if they start at the same point, the one in the inner lane will be ahead of the one in the outer lane as the inner lane is closer to the centre of the circle that defines the bend. The angular velocity of the inner sprinter is larger than that of the outer sprinter, so the group of sprinters 'rotates' differentially. (Of course, to compensate for this the lanes are staggered so that the sprinter in the inner lane starts further back than the one in the outer lane.)

Question 24

The kinematic viscosity ν is related to the dynamical viscosity η via the density ρ of the material: $\eta = \nu\rho$. This expresses just two different definitions of the same physical quantity. It could be said that the kinematic viscosity measures the viscosity relative to the density of the medium.

Question 25

As $\nu \approx \upsilon\lambda$ the kinematic viscosity must have the unit $(\text{m}\,\text{s}^{-1}) \times \text{m}$, i.e. $\text{m}^2\,\text{s}^{-1}$. The dynamical viscosity is just the product $\nu\rho$, hence must have the unit $(\text{m}^2\,\text{s}^{-1}) \times (\text{kg}\,\text{m}^{-3}) = \text{kg}\,\text{s}^{-1}\,\text{m}^{-1}$. This is consistent with Equation 30 where the right-hand side has the unit $\text{kg}\,(\text{s}\,\text{m})^{-1} \times (\text{m}\,\text{s}^{-1})/\text{m} = \text{kg}\,\text{s}^{-2}\,\text{m}^{-1}$, which is the same as for the left-hand side (force divided by area has the unit $(\text{kg}\,\text{m}\,\text{s}^{-2})/\text{m}^2 = \text{kg}\,\text{s}^{-2}\,\text{m}^{-1}$).

Question 26

Equation 36 reads $G = 2\pi R\nu\Sigma R^2 \dfrac{\partial\Omega}{\partial R}$.

The unit of the right-hand side is

$\text{m} \times (\text{m}\,\text{s}^{-1} \times \text{m}) \times (\text{kg}\,\text{m}^{-2}) \times \text{m}^2 \times (\text{s}^{-1}\,\text{m}^{-1})$

Collecting terms this is $\text{m}^{5-3}\,\text{s}^{-2}\,\text{kg} = \text{m}^2\,\text{s}^{-2}\,\text{kg}$.

Now we have torque = force × distance, so the corresponding unit is

$\text{N} \times \text{m} = (\text{kg}\,\text{m}\,\text{s}^{-2}) \times \text{m} = \text{kg}\,\text{m}^2\,\text{s}^{-2}$, as above.

Question 27

The expression for the viscous dissipation rate is

$$D(R) = \frac{1}{2}\nu\Sigma\left(R\frac{\partial\Omega}{\partial R}\right)^2$$

Hence the dimension of viscous dissipation is $\mathrm{m^2\,s^{-1} \times kg\,m^{-2} \times m^2 \times (s^{-1}\,m^{-1})^2}$, i.e. $\mathrm{m^{4-4}\,s^{-1-2}\,kg}$, therefore the unit of viscous dissipation is $\mathrm{kg\,s^{-3}}$.

Another way of obtaining this result is by noting that, physically, the dissipation rate measures the conversion of energy per time interval and per surface area. As energy is force × distance the unit of energy therefore is $\mathrm{kg\,m^2\,s^{-2}}$, and hence the unit of the dissipation rate is $\mathrm{(kg\,m^2\,s^{-2})\,s^{-1}\,m^{-2} = kg\,s^{-3}}$.

Question 28

Equation 37 gives the viscous dissipation rate as

$$D(R) = \frac{1}{2}\nu\Sigma\left(R\frac{\partial\Omega}{\partial R}\right)^2$$

This can be simplified if the angular speed Ω is a known function of R. In the case of Keplerian motion the angular speed is given by

$$\Omega = \left(\frac{GM}{R^3}\right)^{1/2}$$

So the radius derivative is

$$\frac{\mathrm{d}\Omega}{\mathrm{d}R} = (GM)^{1/2}\left(-\frac{3}{2}R^{-5/2}\right)$$

Inserting this into Equation 37 gives

$$D(R) = \frac{1}{2}\nu\Sigma R^2 GM\left(-\frac{3}{2}\right)^2 R^{-5}$$

Hence $$D(R) = \frac{9}{8}\nu\Sigma\frac{GM}{R^3}$$

This is FKR Equation 4.30.

Question29

The α-viscosity $\nu = \alpha H c_{\mathrm{s}}$ is an attempt to quantify the viscosity in accretion discs. As the source for this viscosity is of unknown origin, the expression for it is guided by the idea of a turbulent viscosity. If turbulent eddies with a typical size b move with a typical speed υ they give rise to a viscosity $\nu = b\upsilon$. As the turbulent velocity is limited by the sound speed c_{s} and the eddy size b by the disc height H, this viscosity can also be written as $\nu = \alpha H c_{\mathrm{s}}$, where α is a dimensionless constant smaller than unity. The ignorance of the value of α reflects our ignorance of the viscosity mechanism

Question 30

Equation 44 describes the conservation of angular momentum in the disc:

$$R\frac{\partial}{\partial t}(\Sigma R^2\Omega) + \frac{\partial}{\partial R}(R\Sigma v_{\mathrm{R}} R^2\Omega) = \frac{1}{2\pi}\frac{\partial G}{\partial R}$$

The two terms on the left-hand side describe the angular momentum balance when $\partial G/\partial R = 0$, i.e. in the absence of the so-called source term on the right-hand side. In this case the angular momentum J of a disc ring between radii R and $R + \Delta R$ changes only if there is an imbalance between the angular momentum that flows into the ring via the mass that flows into the ring, and the angular momentum leaving the ring via the mass flowing out of the ring. We find the flow rate of angular momentum at radius R by multiplying the mass flow rate $\dot{M}$ by the specific angular momentum this mass has. The specific angular momentum is just $R^2\Omega$, and $\dot{M}$ is given by Equation 42 as

$$\dot{M}(R,t) = -2\pi R v_{\mathrm{R}}\Sigma$$

So the local flow rate of angular momentum is just

$$\dot{J} = -2\pi R v_{\mathrm{R}}\Sigma R^2\Omega$$

To work out the net change ΔJ of the angular momentum J of the disc ring due to this mass flow in a small time interval Δt we take the difference between the local flow rate at $R + \Delta R$ and R, and multiply it by Δt. This can be written as

$$\Delta J = \left[\dot{J}(R+\Delta R,\, t) - \dot{J}(R,\, t)\right] \times \Delta t \approx \Delta R\frac{\partial \dot{J}}{\partial R} \times \Delta t$$

Hence $$\frac{\Delta J}{\Delta t} = \Delta R\frac{\partial \dot{J}}{\partial R} = -\Delta R\frac{\partial(2\pi R v_{\mathrm{R}}\Sigma R^2\Omega)}{\partial R}$$

This beomes $$\frac{\Delta J}{\Delta t} = -2\pi\Delta R\frac{\partial(v_{\mathrm{R}} R\Sigma R^2\Omega)}{\partial R} \qquad (93)$$

On the other hand, the total angular momentum J in the disc ring is J = mass in the ring × specific angular momentum = $2\pi R\Delta R \times \Sigma \times R^2\Omega$. Hence the time derivative of J can be written as

$$\frac{\partial J}{\partial t} = 2\pi R\Delta R\frac{\partial}{\partial t}(\Sigma R^2\Omega) \qquad (94)$$

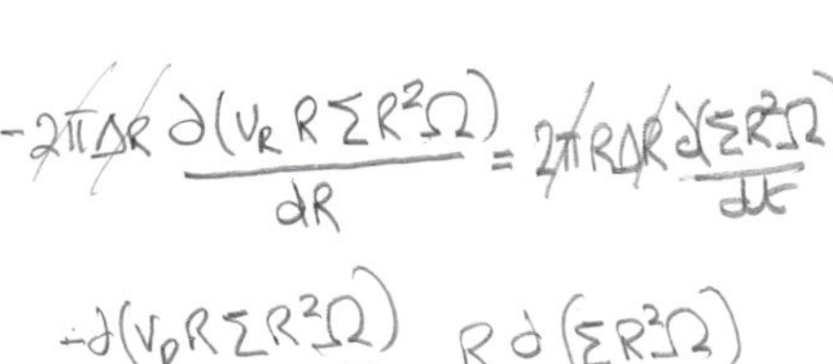

Note that R and ΔR are not affected by the partial derivative with respect to t, as by definition this has to be taken for fixed R. For small time intervals Δt, the expression $\Delta J/\Delta t$ in Equation 93 becomes the derivative $\partial J/\partial t$. Equating the right-hand side of Equation 94 with the right-hand side of Equation 93, and dividing by $2\pi\Delta R$, finally reproduces the first two terms in Equation 44.

Question 31

Equation 49 reads

$$R\frac{\partial \Sigma}{\partial t} = -\frac{\partial}{\partial R}\left[\frac{1}{2\pi\left(\frac{\partial}{\partial R}(R^2\Omega)\right)}\frac{\partial G}{\partial R}\right]$$

We are asked to substitute the Keplerian value $\Omega(R) = (GM/R^3)^{1/2}$ for the angular speed, and the expression

$$G = 2\pi R \nu \Sigma R^2 \frac{\partial \Omega}{\partial R}$$

(FKR Equation 5.5) for the viscous torque. Once done, this should reproduce Equation 45.

The way to proceed is to first calculate the term $\partial G/\partial R$ and then the term $\partial(R^2\Omega)/\partial R$. Both of these terms appear on the right-hand side of Equation 49.

Let us start with $\partial G/\partial R$. First we need to calculate G. In order to do so we need to work out $\partial\Omega/\partial R$:

$$\frac{\partial \Omega}{\partial R} = \frac{\partial}{\partial R}\left(\frac{(GM)^{1/2}}{R^{3/2}}\right) = -\frac{3}{2}\frac{(GM)^{1/2}}{R^{5/2}}$$

Hence G can be written as

$$G = 2\pi R \nu \Sigma R^2 \times \left(-\frac{3}{2}\frac{(GM)^{1/2}}{R^{5/2}}\right)$$

so that $\quad G = -3\pi\nu\Sigma(GMR)^{1/2}$

The derivative of G is therefore

$$\frac{\partial G}{\partial R} = -3\pi(GM)^{1/2}\frac{\partial}{\partial R}(\nu\Sigma R^{1/2}) \tag{95}$$

Now let us turn to $\partial(R^2\Omega)/\partial R$, the other term in Equation 49. We have

$$\frac{\partial}{\partial R}(R^2\Omega) = \frac{\partial}{\partial R}\left(R^2\frac{(GM)^{1/2}}{R^{3/2}}\right) = \frac{1}{2}(GM)^{1/2}R^{-1/2}$$

i.e.

$$\frac{\partial}{\partial R}(R^2\Omega) = \frac{1}{2}\left(\frac{GM}{R}\right)^{1/2} \tag{96}$$

Substituting Equations 96 and 95 into Equation 49, the square bracket on the right-hand side then reads

$$\left[\frac{1}{2\pi\left(\frac{\partial}{\partial R}(R^2\Omega)\right)}\frac{\partial G}{\partial R}\right] = \frac{1}{2\pi}\left(\left(\frac{1}{2}\right)\left(\frac{GM}{R}\right)^{1/2}\right)^{-1}(-3\pi)(GM)^{1/2}\frac{\partial}{\partial R}(\nu\Sigma R^{1/2})$$

By collecting terms this simplifies considerably to

$$\left[\frac{1}{2\pi\left(\frac{\partial}{\partial R}(R^2\Omega)\right)}\frac{\partial G}{\partial R}\right] = -3R^{1/2}\frac{\partial}{\partial R}(\nu\Sigma R^{1/2}) \tag{97}$$

If we replace the square bracket on the right-hand side of Equation 49 by this expression we obtain the equation

$$R\frac{\partial \Sigma}{\partial t} = -\frac{\partial}{\partial R}\left[-3R^{1/2}\frac{\partial}{\partial R}(\nu\Sigma R^{1/2})\right]$$

which can be rearranged to give

$$\frac{\partial \Sigma}{\partial t} = \frac{3}{R}\frac{\partial}{\partial R}\left[R^{1/2}\frac{\partial}{\partial R}(\nu\Sigma R^{1/2})\right]$$

i.e. Equation 45!

Question 32

We are asked to derive FKR Equation 5.9 for the radial drift velocity from Equation 48. Equation 48 reads

$$v_{\mathrm{R}} = \frac{1}{2\pi}\frac{1}{R\Sigma\dfrac{\partial}{\partial R}(R^2\Omega)}\frac{\partial G}{\partial R}$$

You may notice that this task requires the same steps as the solution to Question 31. In particular, the right-hand side of Equation 48 is just the same as Equation 97, divided by $R\Sigma$. If you do not see that this is the case, read now the suggested solution to Question 31.

Inserting Equation 97 into Equation 48 gives

$$v_{\mathrm{R}} = \frac{1}{R\Sigma}(-3R^{1/2})\frac{\partial}{\partial R}(\nu\Sigma R^{1/2})$$

which does indeed simplify to FKR Equation 5.9:

$$v_{\mathrm{R}} = \frac{-3}{\Sigma R^{1/2}}\frac{\partial}{\partial R}(\nu\Sigma R^{1/2})$$

Question 33

(a) Initially, outer parts move out, inner parts move in. (b) The radius where v_{R} changes sign separates the two regimes. (c) With time, the radius where v_{R} changes sign moves out, so at a given radius $R > R_{\mathrm{circ}}$ matter first moves out, but then in again, losing angular momentum to mass that is located yet further out.
(d) Eventually, almost all of the mass has accreted to the centre, with all the angular momentum carried to very large radii by a very small fraction of the mass.

Question 34

Equation 57 is $R\Sigma v_{\mathrm{R}} R^2\Omega = \dfrac{G}{2\pi} + \dfrac{C}{2\pi}$ and the standard form for G is $G = 2\pi R\nu\Sigma R^2\dfrac{\partial\Omega}{\partial R}$.

Inserting this expression for G in Equation 57 gives

$$R\Sigma v_{\mathrm{R}} R^2\Omega = \frac{2\pi R\nu\Sigma R^2}{2\pi}\frac{\partial\Omega}{\partial R} + \frac{C}{2\pi}$$

Simplifying and dividing both sides by R^3 gives

$$\Sigma v_{\mathrm{R}}\Omega = \nu\Sigma\frac{\partial\Omega}{\partial R} + \frac{C}{2\pi R^3}$$

Solving for the term that contains $\partial\Omega/\partial R = \Omega'$ gives

$$-\nu\Sigma\Omega' = \Sigma(-v_{\mathrm{R}})\Omega + \frac{C}{2\pi R^3}$$

which is indeed FKR Equation 5.15.

Question 35

(a) The Keplerian angular speed at radius $R = R_* + b$ is

$$\Omega = \left(\frac{GM}{(R_* + b)^3}\right)^{1/2} = \left(\frac{GM}{R_*^3\left(1+\dfrac{b}{R_*}\right)^3}\right)^{1/2} = \left(\frac{GM}{R_*^3}\right)^{1/2}\left(1+\frac{b}{R_*}\right)^{-3/2}$$

(b) Using the Taylor expansion for the term $(1 + (b/R_*))^{-3/2}$ gives

$$\left(1+\frac{b}{R_*}\right)^{-3/2} \approx 1 - \frac{3}{2}\frac{b}{R_*} + \frac{3}{4}\left(\frac{b}{R_*}\right)^2 + \ldots$$

where the dots (...) denote terms with higher powers b/R_*.

(c) Comparison with FKR Equation 5.17 shows that FKR combines the terms

$$-\frac{3}{2}\frac{b}{R_*} + \frac{3}{4}\left(\frac{b}{R_*}\right)^2 + \ldots$$

into $O(b/R_*)$, consistent with the notation for $O(x)$ (see the box on 'Order'). As b/R_* is small, these terms can be neglected.

Question 36

(a) and (b) The viscous time is $t_{\text{visc}} \sim R^2/\nu$, while the radial drift velocity is $\upsilon_R \sim \nu/R$ (Equations 53 and 54). So we can write $\nu \sim R^2/t_{\text{visc}}$, and also $\nu \sim \upsilon_R R$, for the viscosity. Using this in the product $\nu\Sigma$, shows that $\nu\Sigma$ can be alternatively written as

$$\nu\Sigma \approx \frac{\Sigma R^2}{t_{\text{visc}}} \quad \text{or as} \quad \nu\Sigma \approx \Sigma R \upsilon_R$$

(a) From the first version it is immediately obvious why $\nu\Sigma$ is proportional to the mass accretion rate. ΣR^2 has the dimension of mass; in fact it is roughly the total disc mass, and t_{visc} is the typical time of radial diffusion of mass through the disc.

(b) The second form is reminiscent of Equation 42 by which we defined $\dot{M}$ to begin with.

Question 37

FKR Equation 4.30 reads

$$D(R) = \frac{9}{8}\nu\Sigma\frac{GM}{R^3}$$

Inserting the steady-state relation Equation 61

$$\nu\Sigma = \frac{\dot{M}}{3\pi}\left[1-\left(\frac{R_*}{R}\right)^{1/2}\right]$$

for $\nu\Sigma$ in FKR Equation 4.30 gives

$$D(R) = \frac{9}{8}\frac{\dot{M}}{3\pi}\left[1-\left(\frac{R_*}{R}\right)^{1/2}\right]\frac{GM}{R^3}$$

which simplifies to

$$D(R) = \frac{3GM\dot{M}}{8\pi R^3}\left[1-\left(\frac{R_*}{R}\right)^{1/2}\right]$$

This is just FKR Equation 5.20.

Question 38

Equation 63 describes the luminosity of a disc ring with inner radius R_1 and outer radius R_2:

$$L(R_1,\, R_2) = \frac{3GM\dot{M}}{2}\left\{\frac{1}{R_1}\left[1-\frac{2}{3}\left(\frac{R_*}{R_1}\right)^{1/2}\right]-\frac{1}{R_2}\left[1-\frac{2}{3}\left(\frac{R_*}{R_2}\right)^{1/2}\right]\right\}$$

We obtain the luminosity of the whole disc if we set R_1 equal to the radius of the accreting star, and R_2 to infinity. This is appropriate for an idealized, infinitely extended disc. A real disc in a binary is limited by the size of the Roche lobe of the accreting star. But even in that case the choice $R_2 = \infty$ is usually a rather good approximation, as $R_2 \gg R_1$.

So, with $R_1 = R_*$ and $R_2 \to \infty$ we have

$$L_{\text{disc}} = L(R_*,\, \infty) = \frac{3GM\dot{M}}{2}\left\{\frac{1}{R_*}\left[1-\frac{2}{3}\right]-\frac{1}{\infty}\left[1-\frac{2}{3}\left(\frac{R_*}{\infty}\right)^{1/2}\right]\right\}$$

Clearly the second term in curly brackets is identical to 0 (division by ∞). So

$$L_{\text{disc}} = \frac{3GM\dot{M}}{2}\left\{\frac{1}{3R_*}-0\right\} = \frac{GM\dot{M}}{2R_*}$$

Question 39

(a) Introducing T_* into Equation 65 gives

$$T^4(R) = T_*^4\left(\frac{R_*}{R}\right)^3\left[1-\left(\frac{R_*}{R}\right)^{1/2}\right]$$

or

$$\frac{T^4(R)}{T_*^4} = \left(\frac{R_*}{R}\right)^3\left[1-\left(\frac{R_*}{R}\right)^{1/2}\right]$$

(b) Hence with $y = (T/T_*)^4$ and $x = R/R_*$ we have

$$y(x) = x^{-3}(1-x^{-1/2}) = x^{-3} - x^{-3.5}$$

(c) The maximum value of y is reached at a point x_0 where $\mathrm{d}y/\mathrm{d}x = 0$. As

$$\frac{\mathrm{d}y}{\mathrm{d}x} = -3x^{-4} - (-3.5)x^{-4.5}$$

we have at the maximum

$$0 = -3x_0^{-4} - (-3.5)x_0^{-4.5}$$

Solving for x_0 this becomes

$$3x_0^{-4} = 3.5x_0^{-4.5} \quad \text{or} \quad x_0^{1/2} = 3.5/3$$

hence $x_0 = (7/6)^2$.

(d) The function $(T/T_*)^4$ attains a maximum value at the same radius as T itself does. This radius is $R_0 = R_* x_0$. Hence

$$R_0 = R_* \times \left(\frac{7}{6}\right)^2 = \frac{49}{36} R_*$$

as requested.

(e) Inserting the value for x_0 in the expression for y gives

$$y(x_0) = x_0^{-3} - x_0^{-3.5} = \left(\frac{6}{7}\right)^6 - \left(\frac{6}{7}\right)^7 = \left(\frac{6}{7}\right)^6\left(1 - \frac{6}{7}\right) = \left(\frac{6}{7}\right)^6 \Big/ 7$$

The maximum temperature $T = y(x_0)^{1/4} T_*$ therefore is

$$T = \left[\left(\frac{6}{7}\right)^{3/2} \Big/ 7^{1/4}\right] T_* \approx 0.488 T_*$$

Question 40

According to FKR Equation 5.44 (third line) the characteristic temperature T_* of a neutron star disc is of order 10^7 K. Using this value in Wien's displacement law, $\lambda_{max} T = 2.9 \times 10^{-3}$ m K, gives for the wavelength λ_{max} where the Planck function B_λ is maximal

$$\lambda_{max} = \frac{2.9 \times 10^{-3}\,\text{m K}}{T_*} \cong \frac{2.9 \times 10^{-3}\,\text{m K}}{10^7\,\text{K}} \approx 2.9 \times 10^{-10}\,\text{m} \approx 0.3\,\text{nm}$$

This is in the X-ray region (see Figure 9).

Question 41

Equation 65 is

$$T^4(R) = \frac{3GM\dot{M}}{8\pi R^3 \sigma}\left[1 - \left(\frac{R_*}{R}\right)^{1/2}\right]$$

In the limit $R \to \infty$ we have $R_*/R \to 0$, and hence the term in square brackets becomes unity. Consequently

$$T^4(R) \to \frac{3GM\dot{M}}{8\pi R^3 \sigma}$$

The temperature law in Cheng *et al.* is

$$T^4(R) = \frac{3GM\dot{M}}{8\pi R^3 \sigma}\left(\frac{1}{1-r'} - 2\frac{\sqrt{r'}}{\sqrt{1-r'}}\right)$$

where $\quad r' = 1.5\dfrac{R_s}{R}$

and R_s s the Schwarzschild radius of the black hole. It remains to be shown that the term in brackets tends to 1 in the limit $R \to \infty$.

For $R \to \infty$ we obviously have $r' \to 0$, so that

$$\left(\frac{1}{1-r'} - 2\frac{\sqrt{r'}}{\sqrt{1-r'}}\right) \to \frac{1}{1-0} - 2\frac{\sqrt{0}}{\sqrt{1-0}} = 1$$

Hence both the temperature law given by Equation 65 and by Cheng *et al.* scale as R^{-3} for large R.

Question 42

(a) Figure 4 spans the infrared to the X-ray range of the spectrum, while Figure 5 is a blow-up of Figure 4 and shows the optical/UV range. The new Hubble Space Telescope (HST) observations this paper reports about are in the optical and UV range. In addition, Figure 5 shows earlier observations in the UV by the International Ultraviolet Explorer (IUE) satellite, while Figure 4 collects yet more prior observations that have been published in the literature, including X-ray observations.

(b) The quantity plotted on the ordinate is called AB_ν and is measured in 'mag', i.e. in magnitudes. As explained at the bottom of page 668 (right-hand column), this is called the monochromatic magnitude and is defined as $AB_\nu = -0.5\log_{10} f_\nu - 48.6$. Here f_ν is the observed flux, i.e. the energy a unit surface area receives per second at frequency ν within a frequency interval of width 1 Hz.

(c) The IUE spectra shown in Figure 5 cover the period January–April 1991, while the HST spectrum is from mid-May 1991.

(d) Only the lower (longest) spectrum in Figure 5 has been taken by HST. The data are from May 14/15. This is about 4 months after the outburst which took place in early January 1991.

(e) The variable parameter is the mass accretion rate $\dot{M}$. The figures show model spectra for accretion discs with $\dot{M}$ between $5.6 \times 10^{-7} \mathrm{M}_\odot\,\mathrm{yr}^{-1}$ and $7.7 \times 10^{-9} \mathrm{M}_\odot\,\mathrm{yr}^{-1}$.

(f) Apparently the mass accretion rate decreases with time, as do the disc surface density and temperature. With decreasing surface density the total mass in the disc must decrease as well – the disc drains onto the black hole.

Question 43

UV light is absorbed by the Earth's atmosphere.

Question 44

If the disc is thin we have $H \ll R$. Using Equation 67,

$$H \cong \frac{c_\mathrm{s}}{v_\phi} R$$

this condition can be written as

$$\frac{c_\mathrm{s}}{v_\phi} R \ll R \quad \text{or} \quad c_\mathrm{s} \ll v_\phi$$

But $c_\mathrm{s} \propto T^{1/2}$ (Equation 26; see also Block 1, Section 4.7). Hence if the sound speed is required to be small, the temperature must be small as well.

Question 45

The equation for hydrostatic equilibrium in z-direction is

$$\frac{1}{\rho}\frac{\partial P}{\partial z} = -\frac{GMz}{R^3} \qquad \text{(FKR 5.24) (98)}$$

A suggested solution of this differential equation is

$$\rho(z) = \rho_\mathrm{c} \exp\left(-\frac{z^2}{2H^2}\right) \qquad \text{(FKR 5.33)}$$

with H being a constant (the scale height). To verify that this is indeed a solution for the case of an isothermal atmosphere (T = constant), we note that the pressure and density are related via Equation 25 where $c_s \propto T^{1/2}$ is the constant isothermal sound speed. Hence

$$\frac{1}{\rho}\frac{\partial P}{\partial z} = \frac{1}{\rho}\frac{\partial(c_s^2 \rho)}{\partial z} = \frac{c_s^2}{\rho}\frac{\partial \rho}{\partial z}$$

Using the chain rule we find from FKR Equation 5.33

$$\frac{\partial \rho}{\partial z} = \left(\frac{-2z}{2H^2}\right)\rho_c \exp\left(-\frac{z^2}{2H^2}\right) = -\frac{z}{H^2}\rho$$

Hence $$\frac{1}{\rho}\frac{\partial P}{\partial z} = \frac{c_s^2}{\rho}\left(-\frac{z}{H^2}\rho\right) = -\frac{c_s^2 z}{H^2}$$

Comparing this with Equation 98 shows that the differential equation is fulfilled if

$$-\frac{c_s^2 z}{H^2} = -\frac{GMz}{R^3}$$

i.e. if $$H^2 = \frac{c_s^2}{GM/R^3} \quad \text{or} \quad H = \frac{c_s}{(GM/R)^{1/2}}R$$

The denominator in this last expression is just the Keplerian speed of gas orbiting in the disc at distance R, (Equation 1)

$$v_\phi = \left(\frac{GM}{R}\right)^{1/2}$$

So we finally find that FKR Equation 5.33 solves the vertical hydrostatic balance if the scale height is given as

$$H = \frac{c_s}{v_\phi}R$$

This is indeed just Equation 67.

Question 46

The optical depth has the dimension of $\rho \times \kappa \times z$, i.e. $\text{g cm}^{-3} \times \text{cm}^2\,\text{g}^{-1} \times \text{cm}$. This is dimensionless!

Question 47

The set of equations in FKR Equation 5.41 describes the one-zone model of a steady-state accretion disc. This is to say that the vertical structure is not resolved, but integrated over, and described only by one value of the gas pressure P, density ρ and temperature T, the mid-plane values.

Equation 1 is the definition of the surface density.

Equation 2 relates the disc thickness to the local sound speed and Keplerian speed. It expresses hydrostatic equilibrium in the direction perpendicular to the disc plane.

Equation 3 is the (isothermal) sound speed. The sound speed is used in Equation 2.

Equation 4 is the equation of state. The pressure is given as the sum of the gas pressure of an ideal gas and the radiation pressure.

Equation 5 specifies that in the direction perpendicular to the disc plane energy is transported by radiative diffusion, and assumes that the locally generated heat by viscous dissipation is equal to the rate at which the disc loses energy through the disc surface by radiation.

Equation 6 is the definition of the disc's optical depth in the direction perpendicular to the disc plane. The optical depth is used in Equation 5.

Equation 7 is a direct result of the assumption that the disc is in a steady state, so that the local mass accretion rate $\dot{M}$ is the same everywhere in the disc. This requires that the surface density is inversely proportional to the viscosity.

Equation 8 expresses the fact that the viscosity in the disc could be a complicated function of other variables that describe properties of the disc plasma.

Question 48

The prime observational quantity is the energy flux through the surface area (Equation 37). In a steady-state disc this is *independent* of viscosity (e.g. FKR Equation 5.20). The reason for this is that in a steady state the viscosity *must* adjust itself to obey the equilibrium condition for the surface density and mass transfer rate expressed in Equation 61. So no matter what mechanism is causing the viscosity, the value of $\nu\Sigma$ is always the same.

As a consequence of this, the surface temperature of a steady-state disc, which is – in principle – accessible via the emitted spectrum, is also *independent* of the viscosity (FKR Equation 5.43).

Another quantity that could perhaps be determined observationally is the disc opening angle H/R. How? If the accretor is hot and bright it might irradiate the surface of the donor star. Then the disc casts a shadow on the surface of the donor star. The size of the shadow is a measure for the opening angle. But the second equation in the set FKR Equation 5.49 for Shakura–Sunyaev discs shows that H/R is rather insensitive to α ($H/R \propto \alpha^{-1/10}$, i.e. a change of α by a factor 100 leads to a change in H/R by less than a factor 1.6).

Question 49

The second equation in the set FKR Equation 5.49 is

$$H = 1.7 \times 10^{8} \alpha^{-1/10} \dot{M}_{16}^{3/20} m_1^{-3/8} R_{10}^{9/8} f^{3/5}\ \text{cm}$$

We obtain H/R by dividing both sides of this equation by R.

$$\frac{H}{R} = 1.7 \times 10^{8} \alpha^{-1/10} \dot{M}_{16}^{3/20} m_1^{-3/8} \frac{R_{10}^{9/8}}{R} f^{3/5}\ \text{cm}$$

Now $R_{10} = R/10^{10}\,\text{cm}$, or $R = 10^{10}\,\text{cm} \times R_{10}$, so that

$$\frac{R_{10}^{9/8}}{R} = \frac{R_{10}^{9/8}}{10^{10}\,\text{cm} \times R_{10}} = 10^{-10}\,\text{cm}^{-1} \times R_{10}^{1/8}$$

Hence H/R finally becomes

$$\frac{H}{R} = 1.7 \times 10^{-2} \alpha^{-1/10} \dot{M}_{16}^{3/20} m_1^{-3/8} R_{10}^{1/8} f^{3/5}$$

Question 50

To calculate the disc mass M_{disc} we have to integrate the surface density Σ over the surface of the whole disc. This involves an integral of Σ over all radii R from the inner disc radius R_1 to the outer disc radius R_2:

$$M_{\text{disc}} = 2\pi \int_{R_1}^{R_2} \Sigma R \, \mathrm{d}R$$

(see, for example, Example 28 in Section 3.15 of Block 1). The surface density is a known function of R (Equation 1 of the set FKR Equation 5.49)

$$\Sigma = 5.2\alpha^{-4/5}\dot{M}_{16}^{7/10}m_1^{1/4}R_{10}^{-3/4}f^{14/5}\,\mathrm{g\,cm^{-2}}$$

Following the hint suggested in the question, we set $f = 1$ for simplicity. (See also the comment at the end of the answer.) So we have

$$M_{\text{disc}} = 2\pi \int_{R_1}^{R_2} \Sigma R \, \mathrm{d}R = 2\pi \int_{R_1}^{R_2} (5.2\alpha^{-4/5}\dot{M}_{16}^{7/10}m_1^{1/4}R_{10}^{-3/4}f^{14/5}\mathrm{g\,cm^{-2}}) R \, \mathrm{d}R$$

$$\approx 2\pi \int_{R_1}^{R_2} 5.2\alpha^{-4/5}\dot{M}_{16}^{7/10}m_1^{1/4}R_{10}^{-3/4}\mathrm{g\ cm^{-2}} R \, \mathrm{d}R$$

Noting that $R_{10} = R/10^{10}$ cm and moving all constants in front of the integral this becomes

$$M_{\text{disc}} = 2\pi \times 5.2\alpha^{-4/5}\dot{M}_{16}^{7/10}m_1^{1/4}\ \mathrm{g\,cm^{-2}} \int_{R_1}^{R_2} \left(\frac{10^{10}\ \mathrm{cm}}{R}\right)^{3/4} R \, \mathrm{d}R$$

$$M_{\text{disc}} = 2\pi \times 5.2 \times (10^{10}\ \mathrm{cm})^{3/4}\alpha^{-4/5}\dot{M}_{16}^{7/10}m_1^{1/4}\ \mathrm{g\,cm^{-2}} \int_{R_1}^{R_2} R^{1/4}\,\mathrm{d}R)$$

For the integral we find

$$\int_{R_1}^{R_2} R^{1/4}\,\mathrm{d}R = \left[\tfrac{4}{5}R^{5/4}\right]_{R_1}^{R_2} = \tfrac{4}{5}\left(R_2^{4/5} - R_1^{4/5}\right) = \tfrac{4}{5}R_2^{5/4}$$

as $R_1 = 0$. Using this in Equation 99 we find for the disc mass

$$M_{\text{disc}} = 2\pi \times 5.2 \times (10^{10}\ \mathrm{cm})^{3/4}\alpha^{-4/5}\dot{M}_{16}^{7/10}m_1^{1/4}\ \mathrm{g\,cm^{-2}}\ \tfrac{4}{5}R_2^{5/4}$$

Setting $R_2 = 10^{11}$ cm (as FKR do just before they evaluate FKR Equation 5.51) we obtain

$$M_{\text{disc}} = 2\pi \times 5.2 \times \frac{4}{5} \times (10^{10})^{3/4} \times (10^{11})^{5/4}\alpha^{-4/5}\dot{M}_{16}^{7/10}m_1^{1/4}\ \mathrm{g\,cm^{-2}} \times \mathrm{cm}^{3/4} \times \mathrm{cm}^{5/4}$$

or $$M_{\text{disc}} = (4.65 \times 10^{22}\ \mathrm{g})\alpha^{-4/5}\dot{M}_{16}^{7/10}m_1^{1/4}$$

Noting that $1\mathrm{M}_\odot = 2 \times 10^{33}$ g, this is also

$$M_{\text{disc}} = (2.3 \times 10^{-11}\mathrm{M}_\odot)\alpha^{-4/5}\dot{M}_{16}^{7/10}m_1^{1/4}$$

The difference between this and FKR Equation 5.51 is twofold: FKR chose not to include (or forgot) the factor $m_1^{1/4}$, and the numerical coefficient is rounded up to the next decade – which is sensible since the purpose of this exercise was to calculate an upper limit of the disc mass.

Note: In order to integrate the surface density over R we had to decide what to do with f, which is also a function of R. We chose the easy option and used $f = 1$ to get an approximate result. In fact, the value we obtained is an *upper limit* on the disc mass.

This can be seen as follows. The inequality $f < 1$ holds for all radii, hence also $f^{14/5} < 1$. The significance of this is that, quite generally, if a relation $a(x) < b(x)$ holds for two functions $a(x)$, $b(x)$, then there is a similar relation $\int a\,\mathrm{d}x < \int b\,\mathrm{d}x$ between the integrals of a and b. In particular, the integral of b can be used to calculate an upper limit for the integral of a. In the solution given above we did just that for Σ, by replacing the expression for Σ with one where we set $f = 1$.

Question 51

The α-viscosity is proportional to the product of sound speed and disc scale height, i.e. $\nu = \alpha c_s H$.

According to Equation 67 the scale height is related to the sound speed and the local Keplerian speed via

$$H \approx \frac{c_s}{\upsilon_\phi} R$$

Hence at a given R we have $\nu \propto c_s H \propto c_s^2$.

The isothermal sound speed c_s is proportional to the square root of the temperature (Equation 26), hence $c_s^2 \propto T$. In the one-zone model of the disc the relevant temperature is the mid-plane temperature T_c, hence $\nu \propto T_c$.

Question 52

Cataclysmic variables are ideal laboratories for the study of accretion phenomena. The reasons are manifold – some of them are:

- The mass donor is faint and does not swamp the optical and UV radiation emitted by the accretion flow itself.
- The irradiation of the accretion disc by the hot accreting white dwarf is negligible.
- The size of the orbit is compact enough so that orbital changes can be observed within hours, a convenient timescale for human observers.
- Eclipses and radial velocity studies allow one to map of the accretion flow.
- Major brightness variations of the disc due to thermal and viscous evolution occur on a convenient timescale of weeks to months.

This last point will become clearer once you have studied Section 8.

Question 53

(a) The accretion luminosity is given by Equation 2

$$L_{acc} = \frac{GM\dot{M}}{R_*}$$

We set $\dot{M} = 10^{-9}\mathrm{M}_\odot\,\mathrm{yr}^{-1}$, $M = 1\mathrm{M}_\odot$, and $R = 8.7 \times 10^8$ cm. Then

$$L_{acc} = (6.673 \times 10^{-10}\,\mathrm{dyne\,cm^2\,g^{-2}}) \times (1.99 \times 10^{33}\,\mathrm{g}) \times [10^{-9} \times 1.99 \times 10^{33}\,\mathrm{g} \times (365.25 \times 24 \times 3600\,\mathrm{s})^{-1}] / 8.7 \times 10^8\,\mathrm{cm}$$

$$L_{acc} = 9.6 \times 10^{31}\,\mathrm{dyne\,cm\,s^{-1}} \quad \text{or} \quad L_{acc} \approx 10^{31}\,\mathrm{erg\,s^{-1}}$$

The solar luminosity is $L_\odot \approx 4 \times 10^{33}\,\mathrm{erg\,s^{-1}}$. Hence using the definition for magnitudes (see Block 1, Section 2.7) we have for the difference between the absolute magnitude M_{CV} of the CV and the absolute magnitude M_{Sun} (not to be confused with the solar mass!)

$$M_{\mathrm{CV}} - M_{\mathrm{Sun}} = -2.5\log_{10}\left(\frac{L_{\mathrm{acc}}}{\mathrm{L}_\odot}\right)$$

Hence $$M_{\mathrm{CV}} - 4.83 = -2.5\log_{10}\left(\frac{10^{32}}{4\times 10^{33}}\right)$$

which gives

$$M_{\mathrm{CV}} = 4.83 - 2.5\log_{10}(0.025) = 8.84$$

Using the distance modulus (Equation 53 of Block 1, with $A = 0$)

$$m = M_{\mathrm{CV}} - 5 + 5\log_{10}(d/\mathrm{pc})$$

...e of the CV when it is located at a distance

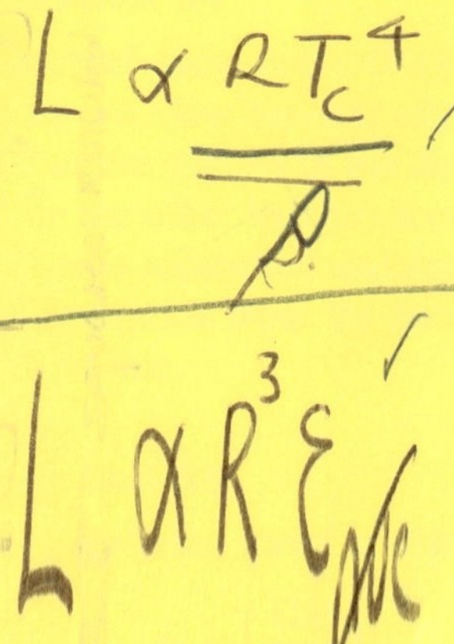

$$\ldots 00) = 8.84 - 5 + 10$$

...$\mathrm{M_\odot\,yr^{-1}}$ this CV is one of the brighter ones anyway, ...hen located fairly close by, at a distance of 1000 pc, ...only just be observable with a 12 inch telescope. ...veys of the sky would of course easily detect such a ...s, but the problem then is to distinguish the CVs in ...ore numerous ordinary stars.

...fect and the rotational motion of the accretion disc. ...or under some angle (but not face-on), then about ...a velocity component away from the observer, ...nent towards the observer. Emission from the ...ne red (long wavelength) part of the double-peaked ...lueshifted and causes the blue part of the double-

...R from the white dwarf with mass M is (Equation 1)

...0^{30} kg), and $R = R_{\mathrm{out}} = 0.5\mathrm{R}_\odot$ ($1\mathrm{R}_\odot = 6.96 \times 10^8$ m)

$$\ldots\left(\frac{\ldots^{11}\,\mathrm{N\,m^2\,kg^{-2}} \times 0.8 \times 1.99 \times 10^{30}\,\mathrm{kg}}{0.5 \times 6.96 \times 10^8\,\mathrm{m}}\right)^{1/2}$$

$$\ldots\mathrm{s^{-1}} = 553\,\mathrm{km\,s^{-1}}$$

For the inner edge of the accretion disc we set $R = R_{in} = 7 \times 10^6$ m. As

$$v_{\phi,\text{in}} = v_{\phi,\text{out}}\left(\frac{R_{out}}{R_{in}}\right)^{1/2}$$

we have with $R_{in} = 2 \times 10^{-2} R_{out}$

$$v_{\phi,\text{in}} = 5.53 \times 10^5\ \text{m s}^{-1} \times \frac{10}{\sqrt{2}} = 3.91 \times 10^6\ \text{m s}^{-1}$$

The Doppler shift is given by

$$\frac{\Delta\lambda}{\lambda} = \frac{v}{c}$$

Hence at the outer edge of the disc the Doppler shift is

$$\Delta\lambda_{out} = \frac{5.53 \times 10^5}{3.0 \times 10^8} \times 656\ \text{nm} = 1.2\ \text{nm}$$

while at the inner edge of the disc the Doppler shift is

$$\Delta\lambda_{in} = \frac{3.91 \times 10^6}{3.0 \times 10^8} \times 656\ \text{nm} = 8.6\ \text{nm}$$

Question 56

The x-axis is the line that connects the two stellar centres, while the y-axis is parallel to the line of sight (if $i = 90°$). The surface pattern arises from lines that connect points in the disc with constant magnitude of the radial velocity, i.e. constant magnitude of the y-component of the orbital velocity v.

Consider now two points, A and B, in the accretion disc that are mirror-symmetric with respect to the y-axis. If point A has coordinates (x_0, y_0), then point B must have coordinates $(-x_0, y_0)$. The symmetry with respect to the y-axis arises because v_y at A has the same magnitude but opposite sign than v_y at B. Therefore, the only difference between a point 'to the left' (A) and 'to the right' (B) of the y-axis is that the plasma to the left is approaching, while the plasma to the right is receding from the observer (if the orbital motion is counter clockwise).

The situation is similar if we consider two points, A and C, in the accretion disc that are mirror-symmetric with respect to the x-axis. As point A has coordinates (x_0, y_0) point C must have coordinates $(x_0, -y_0)$. The symmetry with respect to the x-axis arises because v_y at A has the same magnitude *and* sign as v_y at C! The velocities at A and C differ only in the sign of the x-component of v.

Question 57

The bright spot (hot spot) is the region where the accretion stream from the mass donor star impacts the accretion disc. Kinetic energy of the stream is converted into heat and ultimately radiation, hence the impact region is hotter and more luminous than the rest of the outer disc.

The hot spot appears brightest when it faces the observer, i.e. immediately before phase 0 (in Figure 54 this is at phase 0.875). The binary is in phase 0 when the secondary star is closest to the observer. At the opposite phase, close to phase 0.5, the hot spot is facing away from the observer. Hardly any light from the spot reaches the observer as the disc is in the way. This variable contribution from the hot spot gives rise to an orbital 'hump' in the optical light curve. The hump is most pronounced when the system is seen nearly edge-on. If we see the system face-on, there is no such hump. In this case the hot spot always contributes roughly the same (small) amount to the total light.

Question 58

The 'shadow' in the figure indicates those regions on the accretion disc from where the Earth (i.e. the telescope that collects photons emitted from the disc) cannot be seen because it is obscured by the donor star.

The shadow is long if the inclination is high. In a system seen edge-on (inclination 90°) the shadow formally has an infinite length, while in a system seen face-on (inclination 0°) there is no shadow.

Question 59

The figure shows two panels with five light curves each. The difference between the two panels is that all data shown in the left panel have been obtained in the so-called high state of UU Aqr, while the data in the right panel are from the low state of UU Aqr. The system UU Aqr shows long-term brightness variations; in the high state it is about 0.3 mag brighter than in the low state.

The five different light curves in each panel have been obtained in the five different photometric wavebands (i.e. wavelength regions) U, B, V, R and I (in order of increasing wavelength; see Block 1, Section 2.8).

Question 60

The observed radial temperature profiles are compared with theoretical radial temperature profiles of standard steady-state discs (see Section 5.1). As can be seen from Equation 65 the theoretical temperature profile depends on the mass accretion rate $\dot{M}$, and the mass M and radius R_* of the accreting white dwarf. If M is known or can be estimated, R_* follows from the mass radius relation for white dwarf. Then the value of $\dot{M}$ can be determined by matching up the theoretical and observed temperature profiles.

Additional complications arise because the interstellar medium tends to redden the observed light from the CV, and this can distort the observed temperature profile. Also, sometimes the standard disc model profile is a bad fit to the observed profile. This shows that real discs are not always as simple as our optically thick, geometrically thin steady-state disc models.

Question 61

The answer depends on the precise position of the starspot on the equator, and on the mass ratio of the system. The amplitude of the S-wave is a measure of the distance from the rotational axis of the binary system – not the rotational axis of the secondary star!

For simplicity, consider a binary with mass ratio 1. In this case the centre of mass and the L_1 point coincide. Hence if the starspot on the equator is at the L_1 point it is on the binary axis and would not show an S-wave at all. If, however, the starspot on the equator is on the opposite side of the Roche-lobe filling star, then the distance from the rotational axis is maximal. (Note that the secondary stars is tidally locked to the orbit, i.e. the orbital angular velocity is the same as the rotational angular velocity.)

A starspot on the pole of the secondary has an intermediate distance to the rotational axis. So the amplitude of the corresponding S-wave is larger than the amplitude for starspots in the vicinity of the L_1 point, but certainly smaller than the amplitude of the starspot on the backside of the star.

Question 62

The Keplerian velocity for accretion disc material is

$$v = \left(\frac{GM}{R}\right)^{1/2}$$

i.e. it increases with decreasing R. In a velocity map, the surface brightness of the accretion disc is plotted as a function of the x- and y-component of the velocity v of the emitting material. Hence the rapidly moving material from the inner regions of the accretion disc will appear at large values of v_x and v_y, while the slowly moving material from the outer regions of the accretion disc will appear at small values of v_x and v_y.

Question 63

The innermost disc regions are hottest and therefore responsible for the observed X-ray flux from X-ray binaries. These regions are best seen and most prominent in systems with small orbital inclination, e.g. in systems which are seen face-on or almost face-on. But for eclipses to occur the orbital inclination has to be large, in which case part of the X-ray flux is obscured by the disc itself. A sample of bright X-ray sources on the sky is likely to be biased towards these intrinsically bright X-ray binaries, and hence would show on average a smaller fraction of eclipsing systems than a complete, unbiased sample of X-ray binaries.

Question 64

An X-ray dip is a sudden drop of the X-ray intensity to almost zero flux (Figure 75a). The dips are confined to a narrow phase interval and occur preferentially just before orbital phase 0. The dipping activity can be understood as a total eclipse of a point-like X-ray source (the immediate vicinity of the central compact object) by some obscuring material.

(X-ray dips do not occur if the inclination is very close to 90°, suggesting that in this case the central point-like X-ray source is permanently obscured.)

X-ray dips are different from partial X-ray eclipses which occur when the secondary star moves in front of an X-ray bright source (Figure 75b). Eclipses are centred on phase 0 (= phase 1), and the drop in the X-ray light curve is much shallower than for dips. (Unlike dips, X-ray eclipses occur also in systems with very large orbital inclination.)

The shape of the light curve shows that the X-ray bright source which is being eclipsed by the secondary is *not* point-like, but extended. The X-ray flux from this extended source declines only gradually as the secondary obscures more and more of the X-ray bright layers.

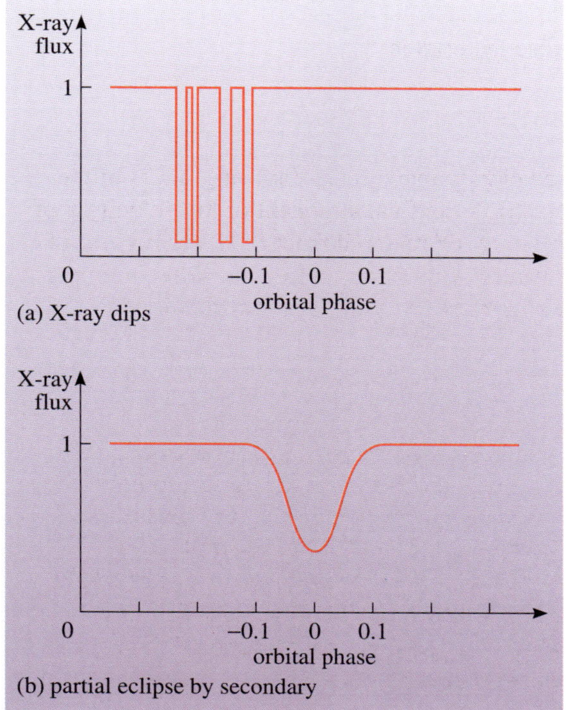

Figure 75 Schematic X-ray light curves. (a) X-ray dips. (b) Partial X-ray eclipse.

Question 65

The obscuring structure that causes X-ray dips must be different from the secondary star (this is also clear from the different phasing of eclipses and dips). In addition, the structure causing the dips must extend higher above the orbital plane than the secondary star does. If this is the case there is a range of inclinations such that, as seen from the observer, the X-ray source can still be in the shadow of this structure, while it is no longer in the shadow of the secondary star.

Question 66

We have the relation $H/R = \tan\delta$ (see Figures 63 and 64). Hence $\delta = 11.3°$.

Question 67

(a) The approximate conservation of magnetic flux means that the product of surface field strength and radius squared, $B_* \times R_*^2$, is roughly constant throughout the life of a star. For the Sun we have $B_* = 1$ G, $R_* = 1\mathrm{R}_\odot$, while for the white dwarf the surface field is B_{WD} and the radius is 8.7×10^8 cm. Hence

$$1\,\mathrm{G} \times (6.96 \times 10^{10}\,\mathrm{cm})^2 = B_{\mathrm{WD}} \times (8.7 \times 10^8\,\mathrm{cm})^2$$

and so
$$B_{\mathrm{WD}} = 1\,\mathrm{G} \times \left(\frac{6.96 \times 10^2}{8.7}\right)^2 = 6.4 \times 10^3\,\mathrm{G}$$

The white dwarf surface magnetic field strength is almost 10^4 G.

(b) A typical red giant radius is several $100\mathrm{R}_\odot$. So compared to the Sun, the surface magnetic field on a red giant is weaker by more than a factor of $100^2 = 10^4$, i.e. a typical surface field strength is less than 10^{-4} G. In that sense red giants are the least magnetic stars.

Question 68

The magnetic moment is given by $\mu = B_* R_*^3$

So for a typical pulsar this is $\mu_{\text{pulsar}} = 10^{12}\,\text{G} \times (10 \times 10^5\,\text{cm})^3$, i.e. $\mu_{\text{pulsar}} = 10^{30}\,\text{G cm}^3$. This is 10^4 times smaller than the magnetic moment of a white dwarf ($10^{34}\,\text{G cm}^3$) in a polar.

Question 69

The Alfvén radius is defined in the context of plasma that accretes spherically symmetrically onto a magnetic accretor. The Alfvén radius is the distance from the accretor where the magnetic pressure just equals the ram pressure of the in-falling material. The significance of this radius is that for radii much smaller than the Alfvén radius the accretion flow is magnetically controlled and will essentially follow the field lines, while for radii much larger than the Alfvén radius magnetic effects on the accretion flow are negligible.

The concept of the spherical Alfvén radius is useful even for very different accretion geometries, such as the accretion via a disc. The critical radius which signals the point where the magnetic field becomes important for the plasma flow is usually of order the Alfvén radius.

Question 70

(a) The combined period distribution (polars and intermediate polars taken together) shows a short-period cut-off, the so-called minimum period. Due to the particular bin width used in the plot, the minimum period appears to be at about 90 min, while the list of individual periods reveals that it is actually at around 80 min. In addition, the combined period distribution shows two peaks, one at periods shorter than 2 hours, the other at periods longer than 3 hours. (In the period distribution of *all* CVs, not just the magnetic CVs you have been looking at, this feature becomes the so-called CV period gap, a dearth of systems in the period range 2–3 hours.) At longer periods the period distribution peters out.

(b) The polars populate predominantly the period range below the period gap, while the intermediate polars (IPs) predominantly populate the period range above the period gap.

Question 71

(a) The suggested criterion for a CV to appear as a polar, i.e. as a system where the magnetic white dwarf rotates synchronously with the orbit, is $R_\text{M} > R_1$.

We use FKR Equation 6.20 ($R_\text{M} = 0.5 \times r_\text{M}$) and Equation 6.18 to calculate R_M, and set $R_1 = 0.5 \times a$.

$$0.5 \times 5.1 \times 10^8 \dot{M}_{16}^{-2/7} m_1^{-1/7} \mu_{30}^{4/7}\,\text{cm} > 0.5 \times a \tag{100}$$

Now we work out what each term in Equation 100 is in our case.

(b) The mass transfer rate in Equation 100 is in units of $10^{16}\,\text{g s}^{-1}$, so we need to convert $\dot{M} = 10^{-10}\text{M}_\odot\,\text{yr}^{-1}$ into cgs units.

$$\begin{aligned}\dot{M}_{16} &= \frac{10^{-10} \times (1.99 \times 10^{33}\,\text{g}) \times (365.25 \times 24 \times 3600\,\text{s})^{-1}}{10^{16}\,\text{g s}^{-1}} \\ &= \frac{6.3 \times 10^{15}\,\text{g s}^{-1}}{10^{16}\,\text{g s}^{-1}} = 0.63\end{aligned}$$

The mass $m_1 = M_1/\mathrm{M}_\odot$ of the white dwarf is given in the question as $M_1 = 0.8\mathrm{M}_\odot$. The corresponding radius of the white dwarf is (Nauenberg's formula)

$$R_{\mathrm{WD}} = 7.83 \times 10^8 \left[\left(\frac{1.44}{m_1} \right)^{2/3} - \left(\frac{m_1}{1.44} \right)^{2/3} \right]^{1/2} \mathrm{cm}$$

$$= 7.83 \times 10^8 \times (1.480 - 0.676)^{1/2} \mathrm{cm} = 7.0 \times 10^8 \mathrm{cm}$$

Therefore the magnetic moment μ_{30}, in units of $10^{30}\,\mathrm{G\,cm^3}$, is

$$\mu_{30} = \frac{B_{\mathrm{WD}} R_{\mathrm{WD}}^3}{10^{30}\,\mathrm{G\,cm^3}} = \frac{10 \times 10^6\,\mathrm{G} \times (7.0 \times 10^8\,\mathrm{cm})^3}{10^{30}\,\mathrm{G\,cm^3}} = 3.4 \times 10^3$$

(c) Finally, to relate a to the orbital period P we use Kepler's law

$$\frac{a^3}{P^2} = \frac{G}{4\pi}(M_1 + M_2)$$

Solving this for a gives

$$a = \left[\frac{G}{4\pi}(M_1 + M_2) \right]^{1/3} P^{2/3}$$

We express the right-hand side in more convenient units

$$a = \left[\frac{6.673 \times 10^{-8}\,\mathrm{dyne\,cm^2\,g^{-2}}}{4\pi^2}(m_1 + m_2) \times 1.99 \times 10^{33}\,\mathrm{g} \right]^{1/3} P_{\mathrm{hr}}^{2/3}(3600\,\mathrm{s})^{2/3}$$

You are told to assume that $m_1 + m_2 = 1.0$, so

$$a = 3.52 \times 10^{10}\, P_{\mathrm{hr}}^{2/3}\,\mathrm{cm} \qquad (101)$$

Using all these in Equation 100 gives

$$0.5 \times 5.1 \times 10^8 \times 0.63^{-2/7} \times 0.8^{-1/7} \times (3.4 \times 10^3)^{4/7}\,\mathrm{cm}$$

$$> 0.5 \times 3.52 \times 10^{10} \times P_{\mathrm{hr}}^{2/3}\,\mathrm{cm}$$

(d) Rearranging gives

$$\frac{5.1 \times 0.63^{-2/7} \times 0.8^{-1/7} \times 3400^{4/7}}{3.52 \times 10^2} > P_{\mathrm{hr}}^{2/3}$$

or

$$P_{\mathrm{hr}}^{2/3} < \frac{5.1}{352} \times \left(\frac{3400^4}{0.63^2 \times 0.8} \right)^{1/7}$$

i.e. $P_{\mathrm{hr}} < 1.78^{3/2}$ and hence $P_{\mathrm{hr}} < 2.4$.

In other words, polars should occur only below a certain orbital period of order 2 hours. This is in broad agreement with observations.

Of course, it is not clear if the simplifying assumption that all polars and IPs have, on average, the same mass transfer rate, white dwarf mass and white dwarf field strength is a good one. There is evidence that at least the field strengths in polars are systematically larger than in IPs. So we should not take the agreement of our estimate with observations as quantitative proof for the assumed synchronization criterion. Nonetheless our result is rather suggestive.

Question 72

(a) The co-rotation radius is the radius where the angular velocity on a Keplerian orbit around the central accreting star equals the rotational angular velocity of this star. Hence, the time it takes to complete one Keplerian orbit at R_Ω is the same as the rotational period of the white dwarf.

(b) According to FKR Equation 6.25 the co-rotation radius is given by

$$R_\Omega = \left(\frac{GMP^2_{\text{spin}}}{4\pi^2}\right)^{1/3} = 1.5\times 10^8\,(P_{\text{spin}}/\text{s})^{2/3}\,m_1^{1/3}\ \text{cm}$$

where M is the mass of the white dwarf. Inserting the values for AE Aquarii gives $R_\Omega = 1.5 \times 10^8\ \text{cm} \times 33^{2/3} \times 0.8^{1/3}$ cm, i.e. $R_\Omega = 9.55 \times 10^8$ cm.

(c) This can be compared with the Roche lobe radius of the white dwarf. From $q = 0.7$ we find $M_2 = 0.7 \times 0.8\text{M}_\odot = 0.5\text{M}_\odot$. We use FKR Equation 4.6, written with the replacement $q \rightarrow 1/q$, to obtain the Roche lobe radius of the primary,

$$\frac{R_1}{a} = \frac{0.49q^{-2/3}}{0.6q^{-2/3} + \log_e(1+q^{-1/3})}$$

Hence $$\frac{R_1}{a} = \frac{0.49\times 0.7^{-2/3}}{0.6\times 0.7^{-2/3} + \log_e(1+0.7^{-1/3})} = \frac{0.6215}{0.7611+0.7544} = 0.410$$

Also, w use Equation 101 from Question 71

$$a = 3.52\times 10^{10}(m_1+m_2)^{1/3}P_{\text{hr}}^{2/3}\ \text{cm}$$

to obtain $R_1 = 0.410\times 3.52\times 10^{10}(m_1+m_2)^{1/3}P_{\text{hr}}^{2/3}\text{cm}$

Hence $R_1 = 0.410\times 3.52\times 10^{10}(0.8+0.5)^{1/3}9.8^{2/3}$ cm

i.e. $R_1 = 7.9 \times 10^{10}$ cm. The co-rotation radius is much smaller than the Roche lobe radius of the primary.

(d) The radius of the white dwarf is 6.3×10^8 cm (using Nauenberg's formula, see also Question 71). So the co-rotation radius is only 9.55/6.3 = 1.5 times larger than the white dwarf radius.

Question 73

In a propeller system the magnetospheric radius is larger than the co-rotation radius.

At the magnetospheric radius R_M the accreting plasma effectively couples to the magnetic field lines which are effectively fixed to the white dwarf and therefore co-rotate with the white dwarf. Hence the plasma is forced to co-rotate with the white dwarf as well when it arrives at R_M. If R_M just equals the co-rotation radius, then the angular velocity the plasma acquires at R_M is just the Keplerian one, i.e. the centrifugal and gravitational forces are in equilibrium. If R_M is larger than the co-rotation radius, then the angular velocity the plasma acquires is larger than the local Keplerian value. (Note that the Keplerian angular velocity drops with distance R from the white dwarf as $R^{-3/2}$.) If this is the case centrifugal forces dominate and the plasma is accelerated away from the white dwarf. This is the propeller effect.

Question 74

Block 1, Equation 140 gives for the cyclotron radius

$$r = \frac{mv}{|q|\,B}$$

As speed = distance/time we have for the time of revolution, T,

$$T = \frac{2\pi r}{v} = \frac{2\pi}{v} \times \frac{mv}{|q|\,B}$$

Hence the corresponding frequency, $\nu = 1/T$, is

$$\nu = \frac{|q|\,B}{2\pi m}$$

For an electron we have $q = e$ and $m = m_e$, so that the cyclotron frequency is

$$\nu_{\rm cyc} = \frac{eB}{2\pi m_e}$$

This is FKR Equation 6.30, with the c in the numerator 'activated', due to the electromagnetic units implicit in writing Block 1, Equation 140 for the cyclotron radius.

We now use $q = e = 1.602 \times 10^{-19}$ C, $m_e = 9.1 \times 10^{-31}$ kg and $B = B_7 \times 10^7$ G $= B_7 \times 10^7 \times 10^{-4}$ T $= B_7 \times 10^3$ T.

As 1 T = 1 N s m^{-1} C^{-1} = 1 kg s^{-1} C^{-1}, the cyclotron frequency is $\nu = 2.8 \times 10^{13}\,B_7$ Hz (where 1 Hz = 1 s^{-1}).

Question 75

(a) HU Aqr has a surface magnetic field strength $B = 35$ MG (see Figure 70). Using FKR Equation 6.30 the cyclotron frequency is

$$\nu_{\rm cyc} = 2.8 \times 10^{13}\left(\frac{B}{10^7\,{\rm G}}\right){\rm Hz}$$

i.e. $\nu_{\rm cyc} = 9.8 \times 10^{13}$ Hz. This is also called the *first* cyclotron harmonic, while the *second* has just twice this frequency:

$$\nu_1 = 9.8 \times 10^{13}\,{\rm Hz} \qquad \nu_2 = 1.96 \times 10^{14}\,{\rm Hz}$$

The corresponding wavelength is $\lambda = c/\nu$, i.e.

$$\lambda_1 = \frac{3 \times 10^8\,{\rm m\,s^{-1}}}{9.8 \times 10^{13}\,{\rm Hz}} = 3.1 \times 10^{-6}\,{\rm m}$$

Hence the first harmonic has wavelength 3.1 μm, while the second harmonic has wavelength $\lambda_2 = \lambda_1/2 = 1.5$ μm.

(b) These wavelengths are in the infrared range (see Figure 9).

Question 76

Equation 82 gives for the cooling rate due to radiation

$$Q^- = \frac{{\rm d}F}{{\rm d}z} \approx \frac{4\sigma T_c^4}{3\kappa\rho H^2}$$

Inserting the Kramers opacity $\propto \rho T_c^{-7/2}$ for κ gives

$$Q^- \propto \frac{T_c^4}{\kappa \rho H^2} \propto \frac{T_c^4}{T_c^{-7/2}\rho^2 H^2} \propto T_c^{15/2}\frac{1}{\Sigma^2}$$

For constant surface density this is proportional to $T_c^{15/2}$.

Question 77

Clearly, the disc is unstable if a temperature perturbation keeps growing once applied. Here we investigate a perturbation towards cooler temperature, i.e. we slightly decrease T_c by a small amount $\Delta T_c < 0$. As a result of this change, the heating and cooling rate will in general no longer balance each other. The difference between Q^+ and Q^- as a result of the perturbation is

$$\Delta(Q^+ - Q^-) = \frac{d(Q^+ - Q^-)}{dT_c} \times \Delta T_c$$

As the first factor on the right-hand side is positive by assumption (Equation 84), while the second factor is negative, we have $\Delta(Q^+ - Q^-) < 0$. Now as

$$\begin{aligned}(Q^+ - Q^-)_{\text{new}} &= \Delta(Q^+ - Q^-) + (Q^+ - Q^-)_{\text{before}} \\ &= \Delta(Q^+ - Q^-) + 0 = \Delta(Q^+ - Q^-) < 0\end{aligned}$$

this says that in the perturbed, cooler state the heating loses out against the cooling, i.e. as a result of the perturbation the disc cools even further. The disc is unstable.

Question 78

The two sets differ only in the Equations 5 and 7.

In the time-dependent set FKR Equation 5.74, Equation 7 is the diffusion equation that describes the viscous evolution of the surface density with time. In the steady-state disc described by the set FKR Equation 5.41 the surface density does not change with time. The diffusion equation is replaced by the steady-state equation which specifies the surface density the disc adopts when a certain mass accretion rate and viscosity is prescribed.

Equation 5 in the time-dependent set is the energy transport equation, stating that the energy generated in a disc ring by viscous dissipation is transported in the z-direction via radiation to the surface of the disc. In the time-independent set this equation has been simplified by using the identity expressed in Equation 7 to replace the product $\nu\Sigma$.

Question 79

We started from an accretion disc in a steady state, i.e. a disc with a surface density profile such that the time-derivative of Σ is zero at all radii. We then applied a perturbation $\Delta\Sigma$ to this surface density profile, i.e. created a new disc with a slightly different radial surface density profile. This new disc is in general not in a steady state, but the new surface density obviously obeys the viscous diffusion equation.

We then simplified the diffusion equation for the new surface density profile considerably by collecting terms that refer to the radial profile of the perturbation alone, and the terms that refer to the original steady-state profile alone. This is not quite straightforward, as the diffusion equation is non-linear. Instead, a new variable is introduced, the product $\mu = \nu\Sigma$, and any change $\Delta\mu$ of μ due to a change $\Delta\Sigma$ in the surface density is approximated by the first (i.e. linear) term in the Taylor expansion, $\Delta\mu \approx \partial(\nu\Sigma)/\partial\Sigma \times \Delta\Sigma$. Using this 'linearization' in the diffusion equation for the new surface density profile, all terms related to the original disc in a steady state cancel (as they trivially fulfil the diffusion equation themselves). The remaining terms form a modified diffusion equation for the change $\Delta\mu$; the modification being the factor $\partial(\nu\Sigma)/\partial\Sigma$ on the right-hand side.

Question 80

A stability analysis studies the reaction of a physical system (e.g. an accretion disc) to perturbations. Initially the system is assumed to be in equilibrium. Then a perturbation is applied to the system, and the reaction of the system is calculated. The stability analysis is said to be linear if the initial perturbations are sufficiently small, so that the resulting change of other quantities can be described by the first (linear) term in the corresponding Taylor expansion with respect to the perturbing quantity. The stability analysis is local if the reaction of the system is studied only in the immediate vicinity of a given point in the system. Therefore, the reaction at this point is assumed to be determined by its immediate vicinity only, not by events far away from this point.

Question 81

FKR Equation 5.19 is also our Equation 61 and expresses the relation between surface density and mass accretion rate in steady-state discs:

$$\nu\Sigma = \frac{\dot{M}}{3\pi}\left[1-\left(\frac{R_*}{R}\right)^{1/2}\right]$$

FKR Equation 5.43, our Equation 65

$$T = \left\{\frac{3GM\dot{M}}{8\pi R^3\sigma}\left[1-\left(\frac{R_*}{R}\right)^{1/2}\right]\right\}^{1/4}$$

gives the surface temperature profile in a steady-state disc. The quantity μ is defined by $\mu = \nu\Sigma$. Using this in FKR Equation 5.19 obviously gives

$$\mu = \frac{\dot{M}}{3\pi}\left[1-\left(\frac{R_*}{R}\right)^{1/2}\right]$$

Solving this for $\dot{M}$ and inserting into FKR Equation 5.43 gives

$$T = \left\{\frac{9GM\mu}{8R^3\sigma}\right\}^{1/4} \quad \text{or} \quad T^4 = \frac{9GM\mu}{8R^3\sigma}$$

At a fixed radius (R = constant) in a steady-state disc the quantity μ is proportional to – and therefore a measure of – the mass accretion rate $\dot{M}$, or surface flux $\propto T^4$. Formally

$$\mu \propto \dot{M} \quad \text{and} \quad \mu \propto T^4$$

Question 82

The heating rate Q^+ describes the energy deposited per unit time and unit volume into the disc by viscous dissipation. This is related to the viscous dissipation rate D per unit surface area as

$$Q^+ \approx \frac{D}{H}$$

where H is the disc thickness (Equation 79). A general expression for D for a Keplerian disc is FKR Equation 4.30,

$$D(R) = \frac{9}{8}\nu\Sigma\frac{GM}{R^3}$$

Hence
$$Q^+ \approx \frac{9}{8}\nu\Sigma\frac{GM}{R^3}\frac{1}{H} \qquad (102)$$

The cooling rate Q^- denotes the rate at which energy is transported away from a unit volume per unit time due to radiative transfer, in the direction perpendicular to the disc plane. One finds

$$Q^- \approx \frac{4\sigma T_\mathrm{c}^4}{3\kappa\rho H^2}$$

(Equation 82). This can be rewritten in terms of the optical depth $\tau = \Sigma\kappa_\mathrm{R}$ (Equation 69), so that

$$\kappa_\mathrm{R}\rho H^2 = \kappa_\mathrm{R} \times (\rho H) \times H = \kappa_\mathrm{R} \times \Sigma \times H = \tau H$$

and hence

$$Q^- \approx \frac{4\sigma T_\mathrm{c}^4}{3\tau H} \qquad (103)$$

If cooling and heating are balanced we have $Q^- = Q^+$.

Using Equations 102 and 103 this is

$$\frac{4\sigma T_\mathrm{c}^4}{3\tau H} = \frac{9}{8}\nu\Sigma\frac{GM}{R^3}\frac{1}{H}$$

which is indeed identical to Equation 5 in FKR Equation 5.74 once H has been cancelled.

COMMENTS ON ACTIVITIES

Completed spreadsheets for all the activities have been installed on your hard disk along with the S381 MM guide. The relevant spreadsheet can be found as *BX_AcY.sdc* where *X* is the block number and *Y* is the activity number, via a link from the multimedia guide.

Activity 2

A completed spreadsheet for this activity may be found by clicking on the Spreadsheets tab from the S381 MM guide.

Activity 4

There are only two images of nova shells in the Image Archive: Nova Cygni 1992, and the recurrent nova T Pyxidis. If you have difficulty finding these images, search using the keyword 'nova shell'.

Activity 11

The two approximations are

$$\frac{R_2}{a} = \frac{0.49q^{2/3}}{0.6q^{2/3} + \log_e(1 + q^{1/3})} \quad \text{and} \quad \frac{R_2}{a} = 0.462\left(\frac{q}{1+q}\right)^{1/3}$$

For $q = 0.5$ the first one gives $R_2/a \approx 0.320\,79$, while the second one gives 0.320 33. The relative deviation is less than 0.14%.
A completed spreadsheet for this activity may be found by clicking on the Spreadsheets tab from the S381 MM guide.

Activity 12

(a) We use the symbol D_{wind} to denote the denominator on the right-hand side of Equation 24 (the so-called stability denominator). To rewrite D_{wind} as a function of mass ratio $q = M_2/M_1$ we first note that $M = M_1 + M_2$. Hence

$$\begin{aligned} D_{\text{wind}} &= \frac{\zeta}{2} + \frac{5}{6} - \frac{M_2}{3M} - \frac{M_2^2}{M_1 M} \\ &= \frac{\zeta}{2} + \frac{5}{6} - \frac{M_2}{3(M_1 + M_2)} - \frac{M_2^2}{M_1(M_1 + M_2)} \end{aligned}$$

Next we divide the numerator and denominator of the third term of D by M_1, and of the fourth term by M_1^2:

$$D_{\text{wind}} = \frac{\zeta}{2} + \frac{5}{6} - \frac{M_2/M_1}{3(M_1 + M_2)/M_1} - \frac{M_2^2/M_1^2}{M_1(M_1 + M_2)/M_1^2}$$

This can be simplified to give

$$D_{\text{wind}} = \frac{\zeta}{2} + \frac{5}{6} - \frac{M_2/M_1}{3(1 + M_2/M_1)} - \frac{M_2^2/M_1^2}{(1 + M_2/M_1)}$$

Introducing $q = M_2/M_1$ gives

$$D_{\text{wind}} = \frac{\zeta}{2} + \frac{5}{6} - \frac{q}{3(1+q)} - \frac{q^2}{(1+q)}$$

The last two terms can be combined:

$$D_{\text{wind}} = \frac{\zeta}{2} + \frac{5}{6} - \frac{q}{3(1+q)} - \frac{3q^2}{3(1+q)} = \frac{\zeta}{2} + \frac{5}{6} - \frac{q+3q^2}{3(1+q)}$$

Factoring out $3q$ and cancelling 3 in the last term we obtain finally

$$D_{\text{wind}} = \frac{\zeta}{2} + \frac{5}{6} - q \times \frac{1/3+q}{1+q}$$

For $\zeta = 1$ this is

$$D_{\text{wind}} = \frac{4}{3} - q \times \frac{1/3+q}{1+q}$$

(b) See the model spreadsheet for a plot of D_{wind} versus q. (Click on the Spreadsheets tab from the S381 MM guide.)

(c) From the plot it can be seen that the denominator D_{wind} is positive for any $q \lesssim 1.75$.

(d) The corresponding expression for conservative mass transfer is much simpler to obtain. From FKR Equation 4.17 (which is valid for $\zeta = 1$) we see that in this case the stability denominator – we call it D_{cons} – is

$$D_{\text{cons}} = \frac{4}{3} - \frac{M_2}{M_1} = \frac{4}{3} - q$$

D_{cons} is positive for any $q \leq \frac{4}{3} \approx 1.3$.

Activity 21

A completed spreadsheet for this activity may be found by clicking on the Spreadsheets tab from the S381 MM guide.

Activity 29

Use Equation 92, $1 M_\odot \text{ yr}^{-1} = 6.3 \times 10^{25} \text{ g s}^{-1}$, to convert the transfer rates into cgs units.

A completed spreadsheet for this activity may be found by clicking on the Spreadsheets tab from the S381 MM guide.

Activity 39

Sample data files with a list of polars and intermediate polars may be found by clicking on the Spreadsheets tab from the S381 MM guide.

Activity 40

A completed spreadsheet for this activity may be found by clicking on the Spreadsheets tab from the S381 MM guide.

ACKNOWLEDGEMENTS

Chris Watson, Graham Wynn, Michael Truss and Richard West supplied some of the animations used in this block. We thank John Barker, Steve Foulkes and Rob Mundin for a critical reading of the manuscript. Dr Hans Ritter's lecture course manuscript was an inspiring source of information for this text. We thank Bart Willems for help with various figures.

Grateful acknowledgement is made to the following sources for permission to reproduce material within this block:

Cover © 2000 Mark A. Garlick

Figure 1 Reproduced by permission of the AAS. Fig.3 from 'Binary-single star scattering. 1. Numerical experiments for equal masses' by Piet Hut and John N. Bahcall, *The Astrophysical Journal*, © The American Astrophysical Society, Vol. 268. pp. 319–341, 1 May 1983; *Figure 3* Courtesy of Spektrum Akademischer Verlag; *Figure 4* Courtesy of Dr James Benson, United States Naval Observatory; *Figure 6* Wolfgang Voges, Max-Planck-Institut fur extraterrestrische Physik, D-85741 Garching, Germany; *Figure 7* Courtesy of Dan Rolfe, University of Leicester; *Figure 10* Courtesy of Mercedes T. Richards and Mark A. Ratliff; *Figure 11* Constructed from observations made by the AAVSO. Courtesy of John Cannizzo; *Figure 12* Bradley, W. Carroll and Ostlie, Dale A. (1996) 'Close binary star systems', *An Introduction to Modern Astrophysics*, Chapter 17, Figure 17.17, The light curve of V1500 Cyg, a fast nova, used by kind permission of Dr. Harold G. Corwin Jr., California Institute of Technology; *Figure 13* © Till Credner, Allthesky.com; *Figure 14 and 15* Rob Hynes, University of Southampton; *Figure 17* Courtesy of Jerry Orosz, University of Utrecht; *Figure 18* Charles, P. A. and Seward, F. D. (1995) 'Low mass X-ray binary stars', *Exploring the X-ray Universe*, Chapter 8, Figure 8.11, used by kind permission of Professor Walter Lewin, Massachusetts Institute of Technology; *Figure 20* Pringle, J. E. and Wade, R. A. (1985) 'Introduction', Interacting Binary Stars (Cambridge Astrophysics Series), Chapter 1, Cambridge University Press; *Figure 22* Charles, P. A. and Seward, F. D. (1995) 'Massive X-ray binary stars', *Exploring the X-Ray Universe*, Chapter 7, Figure 7.22, Cambridge University Press; *Figure 25* Science Photo Library; *Figure 44(b)* Courtesy of Dan Rolfe, University of Leicester; *Figures 50 and 51* Mark A. Garlick; *Figure 52* Gilliland, R. L. (1986) 'WZ Sagittae: Time-resolved spectroscopy during quiescence', *The Astrophysical Journal*, © The American Astrophysical Society, Vol. 301, pp. 252–261, Figure 1, University of Chicago Press; *Figure 53* Horne, K. and Marsh, T. R. (1986) 'Emission line formation in accretion discs', *Monthly Notices of the Royal Astronomical Society*, Vol. 218, pp. 761–773, Figure 1, © Royal Astronomical Society, Blackwell Science Limited; *Figure 54(b)* Courtesy of Dan Rolfe, University of Leicester; *Figure 55(a)* Schoembs, R. *et al.* (1987) 'Simultaneous multicolour photometry of OY Carinae during quiescence', *Astronomy and Astrophysics*, Vol. 181, pp. 50–56. Figure 2, © European Southern Observatory, Springer-Verlag GmbH & Co. KG; *Figure 56* Courtesy of Raymundo Baptista, UFSC, Brazil; *Figure 57* Courtesy of Tom Marsh, University of Southampton; *Figure 58* Steeghs, D. (1999) 'PhD thesis (from University of St Andrews): spiral waves in accretion discs', Figure 2.3, Danny Steeghs, University of Southampton; *Figures 60 and 61* Courtesy of D. Steeghs, University of Southampton; *Figure 62* Charles, P. A. and Seward, F. D. (1995) 'Low mass X-ray binary stars', *Exploring the X-Ray Universe*, Chapter 8, Figure 8.8, Cambridge University Press; *Figure 64* From MNRAS volume 38, p.217, 1999, 'Warped accretion discs and the long periods in X-ray binaries' by Ralph A. M. J. Wijers and J. E. Pringle. By courtesy of Blackwell Publishers; *Figure 65* Courtesy of SOHO/EIT 304A consortium. SOHO is a project between ESA and NASA; *Figure 66* Courtesy of NASA; *Figure 67(a)* Copyright Richard Megna/Fundamental Photos/Science Photo Library; *Figure 70* Courtesy of Axel Schwope, Astrophysikalisches Institut Potsdam.

APPENDIX

A1 SI units and cgs units

The main units used in science are SI (standing for Système International [d'Unités]).

SI base units

Physical quantity	Name of unit	Symbol of unit
length	metre	m
mass	kilogram	kg
time	second	s
electric current	ampere	A
temperature	kelvin	K
luminous intensity	candela	cd
amount of substance	mole	mol

Standard SI multiples and submultiples

Multiple	Prefix	Symbol for prefix	Sub-multiple	Prefix	Symbol for prefix
10^{12}	tera	T	10^{-3}	milli	m
10^{9}	giga	G	10^{-6}	micro	μ
10^{6}	mega	M	10^{-9}	nano	n
10^{3}	kilo	k	10^{-12}	pico	p
10^{0}	–	–	10^{-15}	femto	f

Common SI unit conversions and derived units

Quantity	Unit	Conversion		
speed	$\mathrm{m\,s^{-1}}$			
acceleration	$\mathrm{m\,s^{-2}}$			
angular speed	$\mathrm{rad\,s^{-1}}$			
angular acceleration	$\mathrm{rad\,s^{-2}}$			
linear momentum	$\mathrm{kg\,m\,s^{-1}}$			
angular momentum	$\mathrm{kg\,m^2\,s^{-1}}$			
force	newton (N)	1 N	=	$1\,\mathrm{kg\,m\,s^{-2}}$
energy	joule (J)	1 J	=	$1\,\mathrm{N\,m} = 1\,\mathrm{kg\,m^2\,s^{-2}}$
power	watt (W)	1 W	=	$1\,\mathrm{J\,s^{-1}} = 1\,\mathrm{kg\,m^2\,s^{-3}}$
pressure	pascal (Pa)	1 Pa	=	$1\,\mathrm{N\,m^{-2}} = 1\,\mathrm{kg\,m^{-1}\,s^{-2}}$
frequency	hertz (Hz)	1 Hz	=	$1\,\mathrm{s^{-1}}$
charge	coulomb (C)	1 C	=	$1\,\mathrm{A\,s}$
potential difference	volt (V)	1 V	=	$1\,\mathrm{J\,C^{-1}} = 1\,\mathrm{kg\,m^2\,s^{-3}\,A^{-1}}$
electric field	$\mathrm{N\,C^{-1}}$	$1\,\mathrm{N\,C^{-1}}$	=	$1\,\mathrm{V\,m^{-1}} = 1\,\mathrm{kg\,m\,s^{-3}\,A^{-1}}$
magnetic field	tesla (T)	1 T	=	$1\,\mathrm{N\,s\,m^{-1}\,C^{-1}} = 1\,\mathrm{kg\,s^{-2}\,A^{-1}}$

In astrophysics, you will frequently see some quantities expressed in the cgs system (standing for centimetre, gram, second). The difference here is that the base units for length and mass are the centimetre and gram, rather than the metre and kilogram, where 1 cm = 10^{-2} m and 1 g = 10^{-3} kg.

Use of the cgs system in turn gives rise to different derived units, so you will often see speeds quoted in cm s^{-1} for instance. Three particular derived units that you should be aware of are the cgs units for energy, force and magnetic field, namely the erg, the dyne and the gauss respectively.

The conversions are

energy	1 joule = 10^7 erg	or	1 erg = 10^{-7} joule = 1 g cm^2 s^{-2}
force	1 newton = 10^5 dyne	or	1 dyne = 10^{-5} newton = 1 g cm s^{-2}
magnetic field	1 tesla = 10^4 gauss	or	1 gauss = 10^{-4} tesla

The basic unit of electric charge is also defined somewhat differently in the cgs system. In fact, *two* different definitions are sometimes seen, the simplest of which is the 'emu' system where

1 coulomb = 0.1 emu or 1 emu = 10 coulomb

A2 Useful constants and conversions

Name of constant	Symbol	cgs/emu value	SI value
Fundamental constants			
gravitational constant	G	6.673×10^{-8} dyne cm^2 g^{-2}	6.673×10^{-11} N m^2 kg^{-2}
Boltzmann constant	k	1.381×10^{-16} erg K^{-1}	1.381×10^{-23} J K^{-1}
speed of light in vacuum	c	2.998×10^{10} cm s^{-1}	2.998×10^{8} m s^{-1}
Planck constant	h	6.626×10^{-27} erg s	6.626×10^{-34} J s
	$\hbar = h/2\pi$	1.055×10^{-27} erg s	1.055×10^{-34} J s
fine structure constant	$\alpha = e^2/4\pi\varepsilon_0\hbar c$	1/137.0	1/137.0
Stefan–Boltzman constant	σ	5.671×10^{-5} erg cm^{-2} K^{-4} s^{-1}	5.671×10^{-8} J m^{-2} K^{-4} s^{-1}
Thomson cross-section	σ_e	6.652×10^{-25} cm^2	6.652×10^{-29} m^2
permittivity of free space	ε_0	8.854×10^{-23} cm^{-2} s^2	8.854×10^{-12} C^2 N^{-1} m^{-2}
permeability of free space	μ_0	4π dyne emu^{-2} s^2	$4\pi \times 10^{-7}$ T m A^{-1}
Particle constants			
charge of proton	e	1.602×10^{-20} emu	1.602×10^{-19} C
charge of electron	$-e$	-1.602×10^{-20} emu	-1.602×10^{-19} C
electron rest mass	m_e	9.109×10^{-28} g	9.109×10^{-31} kg
		0.511 MeV/c^2	0.511 MeV/c^2
proton rest mass	m_p	1.673×10^{-24} g	1.673×10^{-27} kg
		938.3 MeV/c^2	938.3 MeV/c^2
neutron rest mass	m_n	1.675×10^{-24} kg	1.675×10^{-27} kg
		939.6 MeV/c^2	939.6 MeV/c^2
atomic mass unit	u or amu	1.661×10^{-24} g	1.661×10^{-27} kg
Astronomical constants			
mass of the Sun	$M_\odot$	1.99×10^{33} g	1.99×10^{30} kg
radius of the Sun	$R_\odot$	6.96×10^{10} cm	6.96×10^{8} m
luminosity of the Sun	$L_\odot$	3.83×10^{33} erg s^{-1}	3.83×10^{26} J s^{-1}

angular measure

1° = 60 arcmin = 3600 arcsec

1° = 0.01745 radian

1 radian = 57.30°

temperature

absolute zero: 0 K = −273.15 °C

0 °C = 273.15 K

energy

1 eV = 1.602×10^{-19} J = 1.602×10^{-12} erg

1 erg = 10^{-7} J = 6.242×10^{11} eV

1 J = 10^{7} erg = 6.242×10^{18} eV

spectral flux density

1 jansky (Jy) = 10^{-26} W m^{-2} Hz^{-1} = 10^{-23} erg s^{-1} cm^{-2} Hz^{-1}

1 W m^{-2} Hz^{-1} = 10^{26} Jy = 10^{3} erg s^{-1} cm^{-2} Hz^{-1}

1 erg s^{-1} cm^{-2} Hz^{-1} = 10^{-3} W m^{-2} Hz^{-1} = 10^{23} Jy

mass-energy equivalence

1 kg = 8.99×10^{16} J/c^2 (c in m s^{-1})

1 kg = 5.61×10^{35} eV/c^2 (c in m s^{-1})

1 g = 8.99×10^{16} erg/c^2 (c in cm s^{-1})

1 g = 5.61×10^{28} eV/c^2 (c in cm s^{-1})

wavelength

1 nanometre (nm) = 10 Å = 10^{-9} m = 10^{-7} cm

1 ångstrom (Å) = 0.1 nm = 10^{-10} m = 10^{-8} cm

distance

1 astronomical unit (AU) = 1.496×10^{11} m = 1.496×10^{13} cm

1 light-year (ly) = 9.461×10^{15} m = 9.461×10^{17} cm = 0.307 pc

1 parsec (pc) = 3.086×10^{16} m = 3.086×10^{18} cm = 3.26 ly

A3 Mathematical signs and symbols

$\equiv$	identical to
$=$	equals
$\approx$	approximately equals
$\sim$	is of order of (i.e. is less than 10 times bigger or smaller than)
$\neq$	is not equal to
$>$	is greater than
$\gg$	is much greater than
$\geq$	is greater than or equal to (i.e. is no less than)
$\gtrsim$	is greater than or of order of
$<$	is less than
$\ll$	is much less than
$\leq$	is less than or equal to (i.e. is no more than)
$\lesssim$	is less than or of order of
$\propto$	is proportional to
∞	infinity
$\sqrt{x}$	the positive square root of x
$\sqrt[n]{x}$	the nth root of x which is equal to $x^{1/n}$
$\pm$	plus and minus the following number
$\mp$	minus and plus, taken in the same order as a preceding $\pm$
Δx	the change in x
$f(x)$	a function f depending on the variable x
$\lvert x \rvert$	the absolute value of a number (i.e. ignoring any – sign)
$\lvert \boldsymbol{a} \rvert$	the magnitude or length of a vector
$\sum_{i=1}^{N} m_i$	the sum of $m_1 + m_2 + m_3 + \cdots + m_N$
$\langle x \rangle$	the average value of x
$\mathrm{d}y/\mathrm{d}t$, y', $\dot{y}$	the derivative of y with respect to t; or the gradient of y versus t
$\mathrm{d}^2y/\mathrm{d}t^2$, y'', $\ddot{y}$	the second derivative of y with respect to t
$\int_{t_A}^{t_B} x(t)\,\mathrm{d}t$	the definite integral of the t-dependent function $x(t)$ with respect to t, evaluated over the interval from $t = t_A$ to $t = t_B$

A4 The Greek alphabet

name	upper case	lower case	name	upper case	lower case
alpha	A	α	nu	N	ν
beta	B	β	xi	Ξ	ξ
gamma	Γ	γ	omicron	O	o
delta	Δ	δ	pi	Π	π
epsilon	E	ε	rho	P	ρ
zeta	Z	ζ	sigma	Σ	σ
eta	H	η	tau	T	τ
theta	Θ	θ, ϑ	upsilon	Y	υ
iota	I	ι	phi	Φ	ϕ, φ
kappa	K	κ	chi	X	χ
lambda	Λ	λ	psi	Ψ	ψ
mu	M	μ	omega	Ω	ω

A5 The name game

Part 1

G gravitational constant

M mass of a body

L_{acc} accretion luminosity

$\dot{M}$ mass accretion rate

R_* stellar radius of the mass accreting star

L_{edd} Eddington luminosity, given by the mass of the accretor

η efficiency, i.e. energy gain divided by input (here for mass accretion)

Part 2

M_1, m_1 mass of the accreting star; m_1 is this mass expressed in solar mass units

M_2, m_2 mass of the star that transfers mass; m_2 is this mass expressed in solar mass units

M mass of the binary, i.e. the sum of donor mass and accretor mass

q mass ratio, donor mass divided by accretor mass

a separation between the two stars in the binary

P_{orb}, P_{hr} orbital period of the binary; P_{hr} is the orbital period in hours

P this is also sometimes used for the orbital period, but most of the time it is used to denote the pressure (*notation ambiguity*!)

ω, $\boldsymbol{\omega}$ ω is the angular speed of the binary components; $\boldsymbol{\omega}$, the angular velocity, is a vector with magnitude ω, pointing in the direction of the rotational axis (if the orbital motion is counter clockwise when seen from above the orbital plane, $\boldsymbol{\omega}$ points towards the observer)

F_c magnitude of the centrifugal force

$\boldsymbol{F}_{coriolis}$ Coriolis force (a vector)

ρ, $\bar{\rho}$ density, and mean density (of a star)

$-\mathrm{d}M_2/\mathrm{d}t = -\dot{M}_2$ mass transfer rate

R_* here: stellar radius of the Roche-lobe filling star; but sometimes this is the radius of the accretor (*notation ambiguity*!)

R_2 Roche lobe radius of the donor star

J magnitude of the orbital angular momentum

$\dot{J}_{\text{sys}}$ systemic orbital angular momentum loss rate; one example is:

$\dot{J}_{\text{GR}}$ orbital angular momentum loss rate due to the emission of gravitational waves

ζ the mass–radius index of the donor star

e eccentricity of an elliptical orbit

Part 3

R, ϕ, z cylindrical coordinates for the accretion disc

$\Omega(R)$ the angular speed of matter in the accretion disc; Ω is a function of R, and usually assumed to be equal to the Keplerian value

σ shear stress

η dynamical viscosity, i.e. shear stress divided by shear strain

ν kinematic viscosity, i.e. dynamical viscosity divided by density

λ some characteristic scale length, e.g. mean free path, or deflection length

c_{s} sound speed

α free parameter in the expression for the α-viscosity

$\Sigma(R)$ (mass) surface density in the accretion disc; Σ is a function of R

$H(R)$ vertical height, or scale height, of the accretion disc; H is a function of R

$G(R)$ magnitude of the viscous torque that an accretion disc ring exerts on its neighbouring disc ring; G is a function of R. Note that elsewhere a common symbol to denote torque is Γ. Note also that G can also just denote the constant of gravity (*notation ambiguity!*)

$D(R)$ viscous dissipation rate (per unit surface area); D is a function of R

Re Reynolds number

Part 4

Listed for completeness only:

(R, t) radius coordinate (in the cylindrical coordinate system) and time, the main independent variables for the one-zone model of accretion discs. If the accretion disc is in a steady state, the dependence of variables on t disappears

$\Sigma(R, t)$ surface density of the accretion disc; Σ is a function of R and t

Important new variables and symbols:

$\upsilon_{\text{R}}(R, t)$ radial drift velocity in the accretion disc; υ_{R} is a function of R and t. $\upsilon_{\text{R}} > 0$ for outward drift

$\dot{M}(R, t)$	local mass accretion rate (mass flow rate) in the disc; $\dot{M}$ is a function of R and t. $\dot{M} > 0$ for net flow inwards
t_{visc}	viscous timescale
b	radial width (thickness) of the boundary layer
R_*	radius of the accretor; R_* is sometimes also used to denote the radius of the donor star (*notation ambiguity!*)
$L(R_1, R_2)$	luminosity of the accretion disc ring between R_1 and R_2
L_{disc}	luminosity of the whole accretion disc
L_{BL}	luminosity of the boundary layer

Symbols of minor importance (notation not necessarily generally used):

l	characteristic radial scale length of the surface density profile $\Sigma(R)$
u	volume density of heat energy
j_u	heat flow density
C	integration constant

Part 5

i	orbital inclination
T	in general: temperature; specifically: accretion disc temperature. Can be a function of R, z and t; in the one-zone model of a steady-state disc of R only
T_*	characteristic accretion disc temperature
$V(R, z)$	gravitational potential of the accretor, in cylindrical coordinates
$g_z(R, z)$	vertical (z) component of the gravitational acceleration, due to the presence of the accretor; g_z is a function of R and z
$\upsilon_{\text{K}}(R)$	Keplerian speed; i.e. orbital speed on a circular Kepler orbit; υ_{K} is a function of R
$\upsilon_\phi(R)$	azimuthal speed of disc matter; usually assumed to be equal to the Keplerian speed; υ_ϕ is a function R
$\rho_{\text{c}}(R)$	mass density in the disc mid-plane; ρ_{c} is a function of R
$F(R, z)$	vertical flux, i.e. energy flow per unit time and unit surface area perpendicular to the disc plane; F is a function of R and z
$Q^+(R, z)$	'heating rate', i.e. rate of energy generation in the disc, per unit volume, due to viscous heating; Q^+ is a function of R and z
$\tau(R, z)$	vertical optical depth; τ is a function of R and z
κ_{R}	Rosseland mean opacity; κ_{R} is a function of density and temperature
$\dot{M}_{16}$	mass accretion rate in units of $10^{16}\,\text{g}\,\text{s}^{-1}$

Variables of minor importance:

ε	energy generation rate (per unit volume) in stars

Part 6

M_{V}	absolute visual magnitude
m_{V}	apparent visual magnitude
χ^2	name of a quantity that is central to a widely used statistical test

δ	tilt angle of a warped disc
ε	another angle characterizing the shape of a warped disc

Part 7

P_{gas}	gas pressure, due to random, thermal motion of gas particles
P_{ram}	ram pressure, due to bulk motion of gas particles
P_{mag}	magnetic pressure
$\boldsymbol{B}$, B	magnetic field strength B, i.e. magnitude of the magnetic field vector $\boldsymbol{B}$
B_*	surface magnetic field strength
R_*	radius of the accretor (*notation ambiguity!*; see also Part 4)
μ	magnetic moment (*notation ambiguity!*; see also Part 8)
μ_0	magnetic permeability of free space
r_M	Alfvén radius (critical radius for spherical accretion)
R_M	magnetospheric radius (critical radius for accretion through a disc)
R_Ω	co-rotation radius (distance from accretor where Keplerian angular speed equals rotational angular speed of accretor)
R_1	Roche lobe radius of the primary
P_{beat}	beat period
P_{spin}	spin period of the magnetic accretor

Part 8

$\mathcal{M}$	Mach number of the accretion flow, $\mathcal{M}$ is a function of R
t_z	vertical sound crossing time; t_z is a function of R
t_ϕ	dynamical timescale (time to complete one Kepler orbit); t_ϕ is a function of R
t_{th}	thermal timescale; function of R
μ	mean molecular weight, i.e. a function of state variables in the disc; *or* the product of viscosity and surface density, hence a function of R and t. In Section 7 this was used to represent the magnetic moment, or the magnetic permeability (very severe *notation ambiguity!*)
σ	Stefan–Boltzmann constant; in Section 3 this was used to represent the shear stress (*notation ambiguity!*)
$Q^+(R, z)$	'heating rate', i.e. rate of energy generation in the disc, per unit volume, due to viscous heating; Q^+ is a function of R and z
$Q^-(R, z)$	'cooling rate', i.e. rate of energy transported away in the direction vertical to the disc plane, per unit volume, due to radiative diffusion (or convection); Q^- is a function of R and z
$T_c(R)$	disc mid-plane temperature; T_c is a function of R
τ	'typical' vertical optical depth of the disc

INDEX

Glossary terms and their page references are printed in **bold**.